Editor
Lorin Klistoff, M.A.

Managing Editor
Karen Goldfluss, M.S. Ed.

Illustrator
Blanca Apodaca

Cover Artist
Brenda DiAntonis

Art Manager
Kevin Barnes

Art Director
CJae Froshay

Imaging
Alfred Lau

Publisher
Mary D. Smith, M.S. Ed.

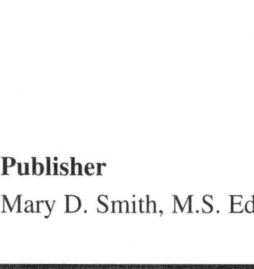

Math in Action: Operations Activities 0–50

Grades 1–2

- Hands-on activities
- Easy-to-use manipulatives
- Lessons & objectives included

Author
Bev Dunbar
(Revised and rewritten by Teacher Created Resources, Inc.)

This edition published by **Teacher Created Resources, Inc.**
6421 Industry Way
Westminster, CA 92683
www.teachercreated.com

ISBN-1-4206-3526-3

©2005 Teacher Created Resources, Inc.
Made in U.S.A.

The classroom teacher may reproduce copies of materials in this book for classroom use only. The reproduction of any part for an entire school or school system is strictly prohibited. No part of this publication may be transmitted, stored, or recorded in any form without written permission from the publisher.

Table of Contents

Introduction 3
How to Use This Book 4
Exploring Addition and Subtraction 6
    How Many Joeys? 7
    Who's Got the Most Cheese? 16
    How Much Fruit? 23
    Eat Your Veggies 29
    Caterpillar Humps 34
    Lily Pad Number Line 39
    Sort the Eggs 43
    Fish Bits 48
    Digit Card Games 50
    Mental Recall Ideas 53
    Addition and Subtraction Check-Up 57
Exploring Multiplication 59
    Multiplication "Active-ities" 60
    Dog Bones 63
    Caterpillars 69
    Smiles 74
    Multiplication Check-Up 79
Exploring Division 81
    Division "Active-ities" 82
    Pirates 83
    Division Review 88
    Division Check-Up 93
Skills Record Sheet 95
Sample Weekly Lesson Plan 96

Introduction

Here are over 50 exciting hands-on, activity-based lessons for teaching addition subtraction, multiplication, and division with whole numbers to 50.

A second book, *Math in Action: Numeration Activities 0–50* (TCR 3525), focuses on developing counting skills and place value concepts. Together, these resources provide you with the practical number ideas you need to keep both you and your young students keenly motivated.

The goal of this series is to actively involve students in mathematics while strengthening their skills and number sense. The activities and lessons are designed to make your life easier, too! With the suggested activities included here, and the numerous reproducible pages and sample programs and Skills Record Sheet, you will find planning your mathematics program for the year much less complicated.

As you explore each topic with your students, you will be able to meet the needs of at least three ability groups, with plenty of opportunities to challenge even your most confident students. For example, in the addition and subtraction section, one group may explore number facts to 10, recording responses with informal word cards. A second group may also explore number facts to 10 but record using the formal word cards. A third group might use the symbol cards and record on paper. One group may even be exploring facts to 20.

Just browse through the relevant section and identify the activities that meet your students' needs. There are enough suggestions in each section to have up to a whole class studying a math topic for at least a week.

Each activity involves active participation and is easy to implement and understand. The themes have immediate appeal for children, who will enjoy manipulating hens and eggs, baby kangaroos, or even pirates as part of their daily math activities.

The activities are designed to maximize the way in which your students build their understanding of numbers and numeration. They are encouraged to think and work mathematically, with an emphasis on mental recall and practical manipulation of objects.

How to Use This Book

❑ Teaching Ideas

Each of the three units has easy-to-implement ideas for several days of teaching! Each activity includes coded skills and grouping strategies to help your planning, programming, and unit assessment. Use each idea for free exploration or guided discovery. You will not run out of exciting teaching ideas for your students!

Coded Skills
(See page 95.)
A = Addition
S = Subtraction
M = Multiplication
D = Division

Grouping Strategies
ONE = individual
PAIR = pair
GROUP = small group
CLASS = whole class

❑ Reproducible Pages

These are the basic resources you need to make the activities come alive in your classroom. There are work cards, flash cards, and spinners for games. Photocopy, color, laminate, and cut out these. Store in suitable containers with lids. Use them for free exploration by groups and individuals or for whole-class demonstrations. Adapt their use to the study of any number. Reuse them later for any operation.

❑ Sets of Four Activity Cards

In addition to free exploration within the activities, each unit includes sets of four activity cards. Use these for guided exploration and to further stimulate and challenge your students. Parent helpers, or capable readers in the class, will also find them useful.

A simple dot code shows how to cater to different ability levels.

How Many Joeys?

✔ Sort some joeys between two kangaroos.
✔ Make up a story to match.
✔ How many joeys does each mother have now?
✔ How many joeys are there altogether?
✔ Record what you discover.

● = match, model, count, and record

●● = guess and check, predict, remember fact, identify more/less

●●● = challenge, use mental skills, remember facts, find alternative solutions

How to Use This Book (cont.)

❏ End-of-Unit Check-Up

The activities in this book encourage open-ended exploration and free recording. However, each unit includes a written worksheet as a check-up to see if the concepts are understood in a more formal way. No reading skills are required.

❏ Skills Record Sheet

The overall objective is to develop knowledge, skills, and understanding of the numbers 0–50 in a variety of fun, child-centered ways.

The specific skills include generating and describing number patterns using various strategies, modelling numbers and number relationships in a variety of ways and using them to solve number problems.

The complete list of skills (page 95) shows you how and when these skills have been reached. Use this checklist to record individual responses during daily activities. (*Note:* The section of each teacher page includes codes such as A1, S4, A3, etc. These codes match the information on the Skills Record Sheet. The record sheet is a quick reference for determining which skills each student has successfully met.)

❏ Sample Weekly Lesson Plan

The Sample Weekly Lesson Plan (page 96) shows you one way to organize a selection of activities from Exploring Multiplication (pages 59–80) as a five-day unit.

Exploring Addition and Subtraction

In this unit, your students will do the following:

❑ Model real-life addition and subtraction stories with concrete materials.
❑ Discover addition and subtraction combinations to 10 or 20 using informal language.
❑ Estimate answers to real-life addition and subtraction story problems.
❑ Record in four stages by using informal word cards, formal word cards, formal symbols, and written symbols.
❑ Learn to use a simple number line to solve addition and subtraction problems.
❑ Recall addition and subtraction facts to 10 or 20 using a wide variety of strategies.

Exploring Addition and Subtraction

How Many Joeys?

Introducing the Activity

Mother kangaroos, in a herd, have to look after the joeys (babies). Sometimes they share this responsibility. We are going to explore all the different ways two mother kangaroos could care for up to 10 joeys between them. Work in a group of up to six students using the kangaroos and joeys in this activity. We will practice recording our discoveries by using word cards for *and*, *take away*, *makes*, and *leaves*.

Skills

- ❏ Demonstrate addition and subtraction facts to 10 with objects (A1, S1).
- ❏ Create and solve story problems (A3, S3).
- ❏ Record addition and subtraction activities using informal word/digit cards (A4, S4).
- ❏ Estimate answers to addition and subtraction problems (A7, S7).

How to Make the Activity

- ❏ Copy, laminate, and cut out the four activity cards (pages 8 and 9).
- ❏ Make six copies of the Mother Kangaroos (page 10) so you have 12 large kangaroo figures in all. Color, laminate, and cut out the cards.
- ❏ Make six copies of the Joeys (page 11). Color, laminate, and cut out to make 60 individual figures.
- ❏ Copy the informal word cards (pages 12 and 13). Make two copies of the digits 0–9 (page 14) for each student. Laminate and cut out along the dashed lines.
- ❏ Place the kangaroos and joeys in a storage container and label clearly. Place the informal word cards in a separate container. Place the digit cards 0–9 in another container.
- ❏ If commercial spinners are not available, copy the spinners (page 15) onto card stock. Cut out, place a large paper clip at the center, and hold a pencil upright with the point pressed firmly at the center, making sure the pencil point sits inside one of the ends of the paper clip. Adjust so that the spinner moves freely.

Extra Materials Required

- ❏ none

Variation

- ❏ Explore addition and subtraction number facts to 20 by copying more joeys (A2, S2). Adapt the ideas on the four activity cards as needed.

How Many Joeys?   Exploring Addition and Subtraction

How Many Joeys?

- ✔ Sort some joeys between two kangaroos.

- ✔ Make up a story to match.

- ✔ How many joeys does each mother have now?

- ✔ How many joeys are there altogether?

- ✔ Record what you discover.

Runaway Joeys

- ✔ Start with ten joeys and one mother kangaroo.

- ✔ Spin a spinner to see how many joeys run away.

- ✔ Count how many joeys the mother has left.

- ✔ Find a way to record your actions.

    Example: | 1 | 0 | take away | 3 | leaves | 7 |

- ✔ Try again with a different number of joeys.

How Many Joeys?   Exploring Addition and Subtraction

How Many Altogether?

✔ Make the first part of a number story with cards.

    Example: ☐3☐ ☐and☐ ☐2☐

✔ Exchange your cards with a friend.

✔ Race to model your new story with kangaroos and joeys.

✔ Try to be the first to record the total number.

    Example: ☐3☐ ☐and☐ ☐2☐ ☐makes☐ ☐5☐

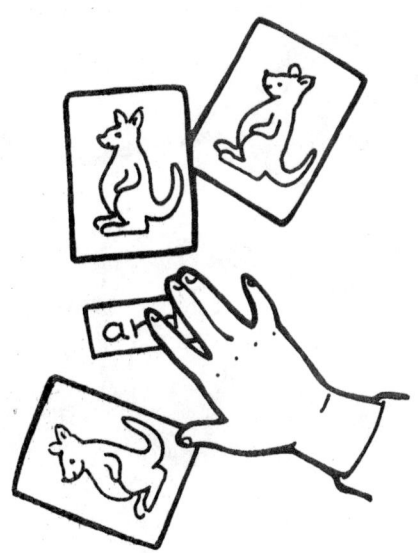

Who's Hiding?

✔ Sort the joeys between two kangaroos.

✔ Record your actions with the small cards.

✔ Hide one mother's joeys and the number card to match.

    Example: ☐4☐ ☐and☐ ☐ ☐ ☐makes☐ ☐7☐

✔ Ask a friend to look at your cards and guess how many joeys are hiding.

✔ Exchange roles.

©Teacher Created Resources, Inc.   #3526 Math in Action

How Many Joeys?   **Exploring Addition and Subtraction**

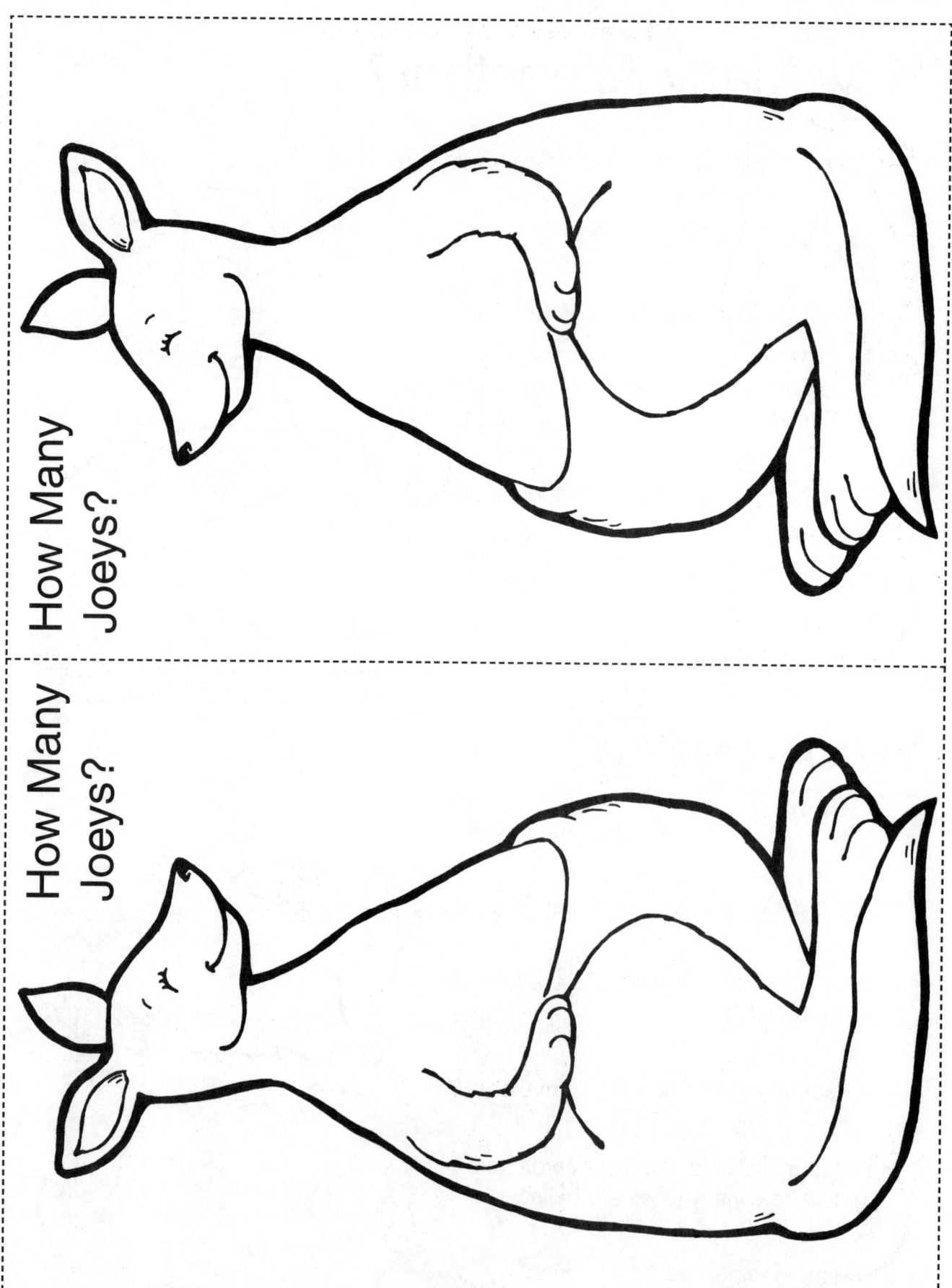

How Many Joeys?

How Many Joeys?

How Many Joeys?   *Exploring Addition and Subtraction*

How Many Joeys?

and	makes
and	makes
and	makes
and	makes
and	makes
and	makes
and	makes

take away	leaves
take away	leaves
take away	leaves
take away	leaves
take away	leaves
take away	leaves
take away	leaves

How Many Joeys? **Exploring Addition and Subtraction**

0	1	2	3
4	5	6	7
8	9	0	1
2	3	4	5
6	7	8	9

How Many Joeys? **Exploring Addition and Subtraction**

#3526 Math in Action ©Teacher Created Resources, Inc.

How Many Joeys? **Exploring Addition and Subtraction**

0–9 Spinner

Directions: Copy and cut out the spinner. Glue or tape it to a piece of cardboard slightly larger than the spinner pattern. To use the spinner, place a large paper clip at the center of the spinner. Hold a pencil upright with the point pressed firmly at the center of the spinner, making sure the pencil point sits inside one of the ends of the paper clip. (See illustration.) To spin the spinner, brush the free end of the paper clip with a finger.

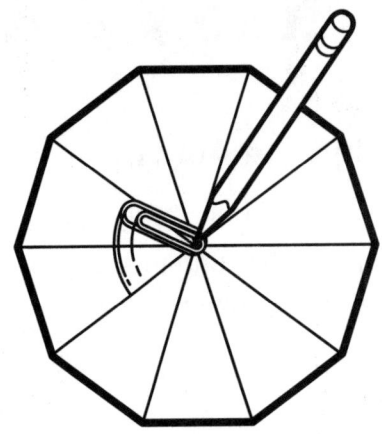

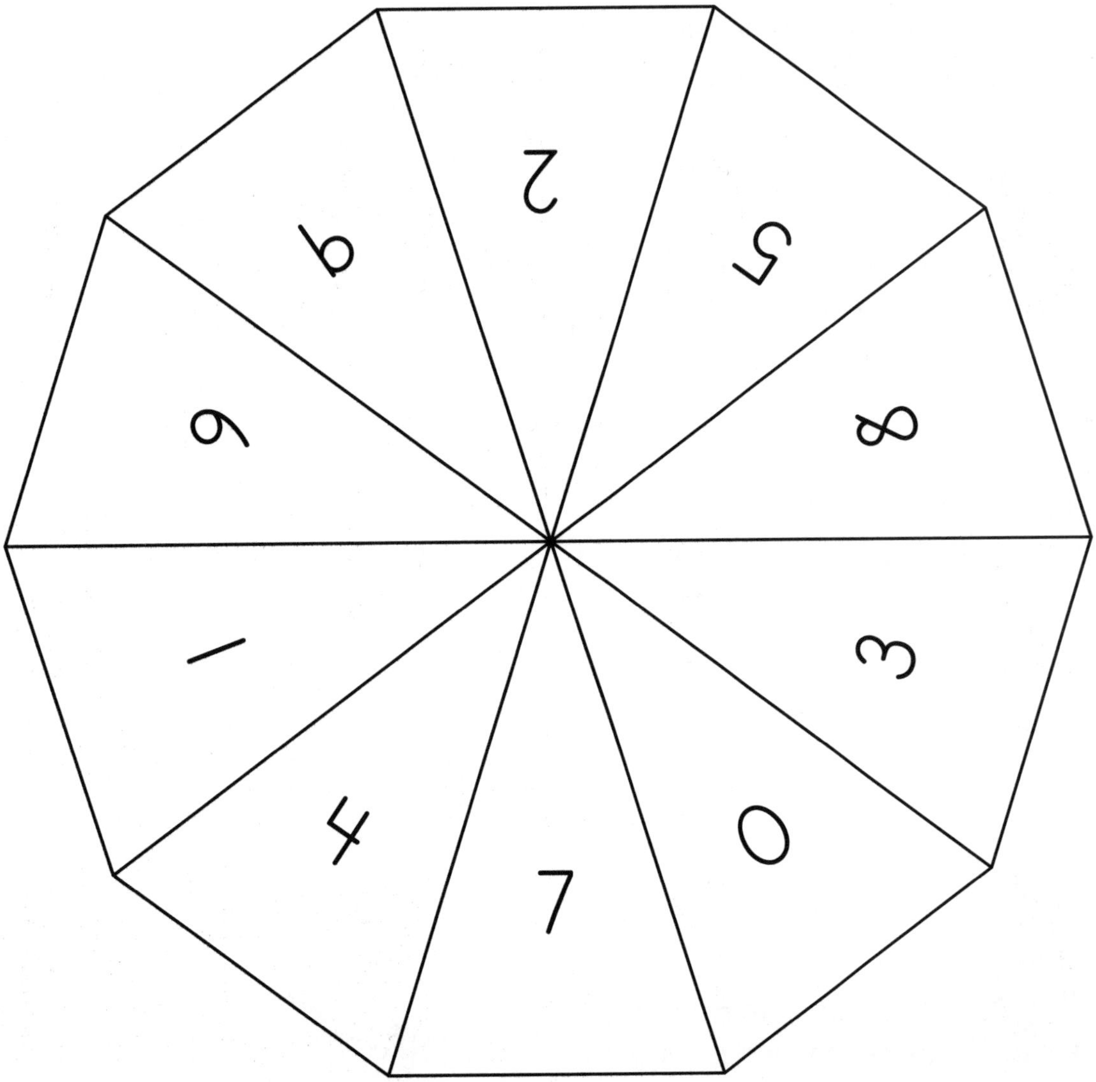

© Teacher Created Resources, Inc. 15 #3526 Math in Action

Who's Got the Most Cheese?

Introducing the Activity

Mice are supposed to love eating cheese. These mice have collected a huge pile to sort between them. Working in a group of up to six students, using the mice and cheese, we will practice using the new terms—*plus, minus, equals* rather than *and, take away, makes,* and *leaves*.

Skills

- ❑ Demonstrate addition and subtraction facts to 10 with objects (A1, S1).
- ❑ Create and solve story problems (A3, S3).
- ❑ Record addition and subtraction activities using formal word/digit cards (A4, S4).
- ❑ Estimate answers to addition and subtraction problems (A7, S7).

How to Make the Activity

- ❑ Copy, laminate, and cut out the four activity cards (pages 17 and 18).
- ❑ Make six copies of the mice (page 19) so that you have 12 mice in all.
  Color, laminate, and cut out as cards.
- ❑ Make six copies of the cheese (page 20). Color, laminate, and cut out as 60 individual pieces.
- ❑ Copy the plus/equals (page 21) or minus/equals (page 22) cards.
  Laminate and cut out along the dashed lines.
- ❑ Place the mice/cheese in a storage container and label clearly.
  Place the word cards in a separate container.

Extra Materials Required

- ❑ two sets of digits 0–9 for each student
  (from previous activity), spinners or dice

Variation

- ❑ Explore addition and subtraction number facts to 20 by copying more cheese (A2, S2). Adapt the ideas on the four activity cards to suit.

Who's Got the Most Cheese?

- ✔ Find three different ways to sort the cheese between two mice.

- ✔ Make up a story to match your actions.

- ✔ Which one has the most cheese?

- ✔ How many pieces are there altogether?

- ✔ Record what you discover.

Race Me

- ✔ Spin the spinner.

  e.g., 7

- ✔ Race to model a story for that number with two mice, cheese, and the small cards.

  e.g., | 8 | minus | 1 | equals | 7 |

Hide the Numbers

✔ Sort the cheese between two mice.

✔ Record with the small cards.

✔ Turn over your number cards.

✔ Ask a friend to look at the mice and cheese. Can your friend guess what your number cards say?

✔ Check and then exchange roles.

Don't Look

✔ Secretly sort some cheese between two mice.

✔ Make a number story to match.

e.g., | 2 | plus | 4 | equals | 6 |

✔ Ask a friend to close his or her eyes and feel how much cheese each mouse has.

✔ Can your friend guess your number story aloud without looking?

✔ Check, then exchange roles.

Who's Got the Most Cheese? **Exploring Addition and Subtraction**

Who's Got the Cheese?

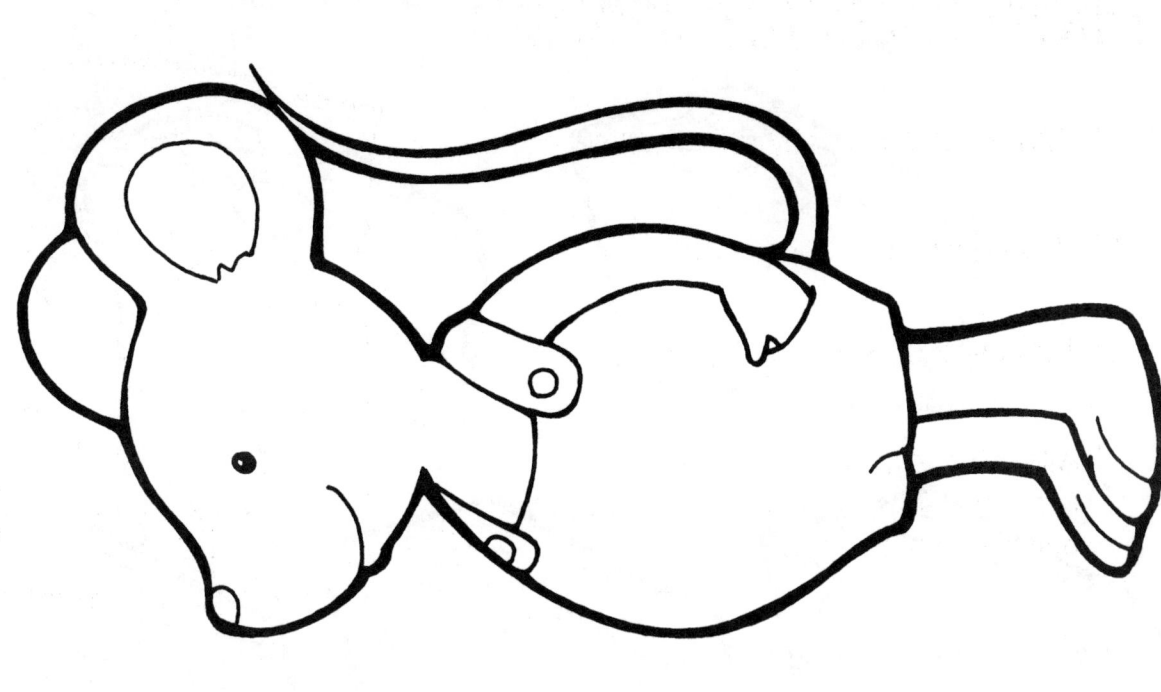

Who's Got the Cheese?

Who's Got the Most Cheese?   Exploring Addition and Subtraction

Who's Got the Cheese?

#3526 Math in Action   ©Teacher Created Resources, Inc.

Who's Got the Most Cheese?   **Exploring Addition and Subtraction**

plus	equals
plus	equals
plus	equals
plus	equals
plus	equals
plus	equals
plus	equals

Who's Got the Most Cheese?   **Exploring Addition and Subtraction**

Who's Got the Most Cheese?   **Exploring Addition and Subtraction**

minus	equals
minus	equals
minus	equals
minus	equals
minus	equals
minus	equals
minus	equals

Who's Got the Most Cheese?   **Exploring Addition and Subtraction**

Exploring Addition and Subtraction

How Much Fruit?

Introducing the Activity

These children love to eat fruit. Try to discover the different ways they can sort up to ten pieces between them. Working in a group of up to six students, using the children and fruit, we will practice recording, using special symbols $\boxed{+}$, $\boxed{-}$, and $\boxed{=}$ instead of our word cards.

Skills

- ❏ Demonstrate addition and subtraction facts to 10 with objects (A1, S1).
- ❏ Create and solve story problems (A3, S3).
- ❏ Record addition and subtraction activities using symbol/digit cards (A5, S5).
- ❏ Estimate answers to addition and subtraction problems (A7, S7).

How to Make the Activity

- ❏ Copy, laminate, and cut out the four activity cards (pages 24 and 25). Make six copies of page 26 so that there are 12 children in all. Color, laminate, and cut out as cards.
- ❏ Make three copies of the fruit (page 27). Color, laminate, and cut out as 75 individual pieces.
- ❏ Copy the +, –, = symbol cards (page 28). Laminate and cut out along the dashed lines.
- ❏ Place the children and fruit cutouts into a storage container and label clearly. Place the symbol cards into a separate container.

Extra Materials Required

- ❏ small plastic fruit counters available from educational suppliers instead of the fruit photocopies (if your budget stretches this far)
- ❏ two sets of digits 0–9 for each student, spinners or dice, three-minute timer

Variation

- ❏ Explore addition and subtraction number facts to 20 by copying more fruit (A2, S2). Adapt the ideas on the four activity cards to suit.

Pieces of Fruit

✔ Make a number story with the small cards.

e.g., | 1 | 0 | - | 6 | = | 4 |

✔ Make up a fruit story to match.

✔ Model your story using the childen and fruit.

✔ How many can you do in three minutes?

Share It

✔ Find different ways the children can have exactly the same number of fruit pieces.

✔ How many pieces are there altogether each time?

✔ Record what you discover.

How Much Fruit?

- ✔ Work with a partner.
- ✔ Turn over a number card each.
- ✔ Give your partner a matching amount of fruit.
- ✔ Who has more? How much more?
- ✔ How many pieces are there altogether?
- ✔ Guess first and then check.
- ✔ Record what you discover.

Fruity Questions

- ✔ Sit with your back to a partner.
- ✔ Secretly give two children some fruit each.
- ✔ Ask each other questions to discover how much fruit is on your partner's two cards.

    e.g., Does the girl have more than the boy?

    Are there more than six pieces?

How Much Fruit? | Exploring Addition and Subtraction

How Much Fruit?

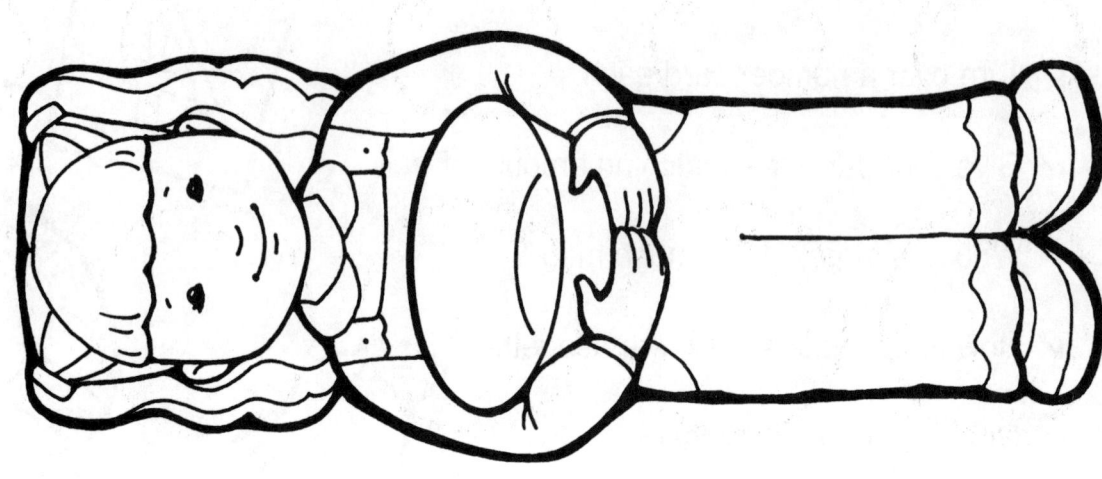

How Much Fruit?

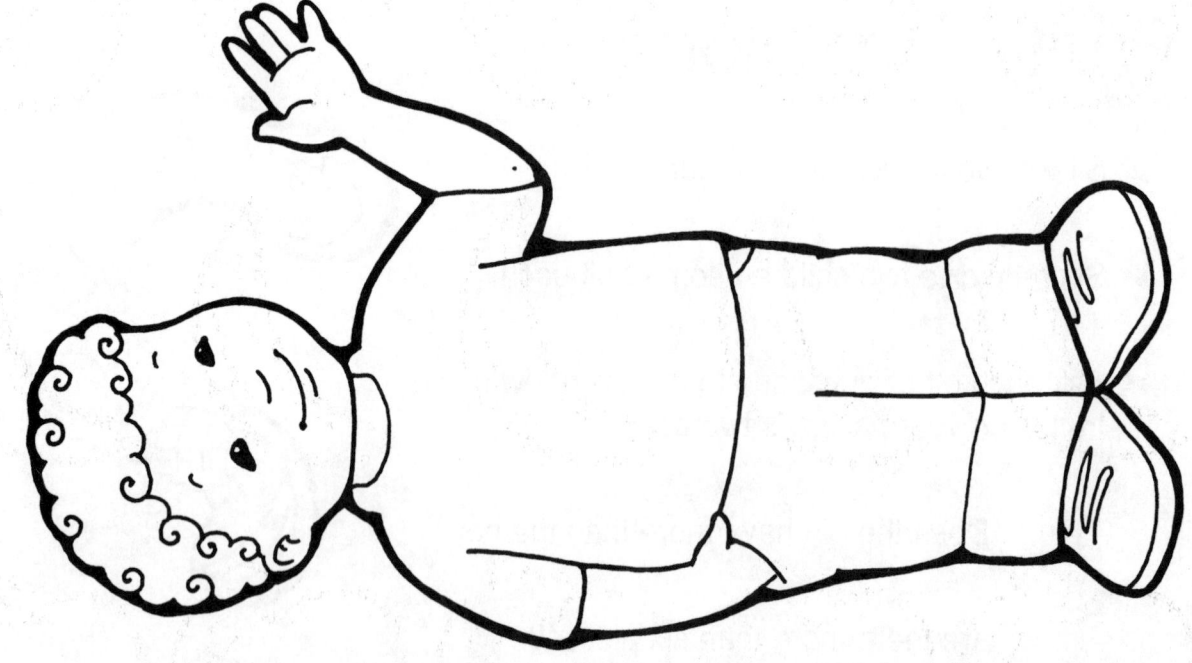

How Much Fruit? | Exploring Addition and Subtraction

#3526 Math in Action | ©Teacher Created Resources, Inc.

How Much Fruit? Exploring Addition and Subtraction

How Much Fruit?

How Much Fruit?   **Exploring Addition and Subtraction**

+	=	−	=
+	=	−	=
+	=	−	=
+	=	−	=
+	=	−	=
+	=	−	=
+	=	−	=

How Much Fruit?   **Exploring Addition and Subtraction**

#3526 Math in Action   28   ©Teacher Created Resources, Inc.

Exploring Addition and Subtraction

Eat Your Veggies

Introducing the Activity

Pretend the counters are different vegetables to eat for your dinner; yummy peas (green), carrots (orange), potatoes (white), tomatoes (red), and so on. Use your imagination to think what all the other colors could be! Working in a group of up to six students using the plates/veggies, we are going to explore ways to join two groups to make up to 20 objects altogether, with symbol cards for recording.

Skills

- ❑ Demonstrate addition and subtraction facts to 20 with objects (A2, S2).
- ❑ Create and solve story problems (A3, S3).
- ❑ Record addition and subtraction activities using symbol/digit cards (A5, S5).
- ❑ Estimate answers to addition and subtraction problems (A7, S7).

How to Make the Activity

- ❑ Copy, laminate, and cut out the four activity cards (pages 30 and 31).
- ❑ Make three copies of the knife/fork/plate cards (page 32) so that you have six cards in all. Color, laminate, and cut out as rectangular cards, or you can use small paper plates.
- ❑ Copy the 10–19 spinners (page 33) onto cardboard. Cut out, place a small skewer through the center. Check for bias.
- ❑ Place the equipment into a storage container. Label clearly.

Extra Materials Required

- ❑ 120 plastic counters (e.g., 15 mm diameter) in a variety of bright colors (Alternatively, plastic 3-D veggie counters are also available from educational suppliers.)
- ❑ two sets of digits 0–9 for each student, symbol (+, –) cards, 0–9 spinners (page 15)

Variations

- ❑ Some children may want to explore combining or removing three groups or more, using up to 20 objects in total.
- ❑ Practice estimating. Take a handful of fruit, guess how many pieces, and then check by counting. How many different ways can you sort these veggies into two groups? Repeat. Do your guesses get closer?
- ❑ Some students may want to explore number combinations to 30 or more.

Eat Your Veggies

✔ Discover different ways to combine veggies on your plate.

✔ Make up stories to match.

✔ Record your actions with the small cards.

e.g., | 1 | 2 | + | 4 | = | 1 | 6 |

Veggie Spin

✔ You have more than 12 veggies on your plate.

✔ Use a 0–9 spinner to show how many pieces you eat.

✔ Guess how many veggies you have on your plate now. Find a way to check.

✔ Record your actions with the small cards.

e.g., | 1 | 3 | – | 3 | = | 1 | 0 |

Make up another story like this.

Eat Your Veggies **Exploring Addition and Subtraction**

Veggie Combo

✔ Spin the 10–19 spinner.

✔ How many different ways can you combine veggies to make this number in three minutes?

    e.g.,   16:   7 peas and 9 carrots

            14 tomatoes and 2 potatoes

✔ Record what you discover.

    e.g., | 7 | + | 9 | = | 1 | 6 |

How Many Veggies Now?

✔ Spin the 0–9 spinner.

✔ Take that many veggies.

✔ Spin the 10–19 spinner.

✔ Guess how many veggies you have to add to make that total altogether. Check.

(Try taking away by using the 10–19 spinner first and then taking away whatever the 0–9 spinner shows.)

Eat Your Veggies Exploring Addition and Subtraction

Spinners
10–19

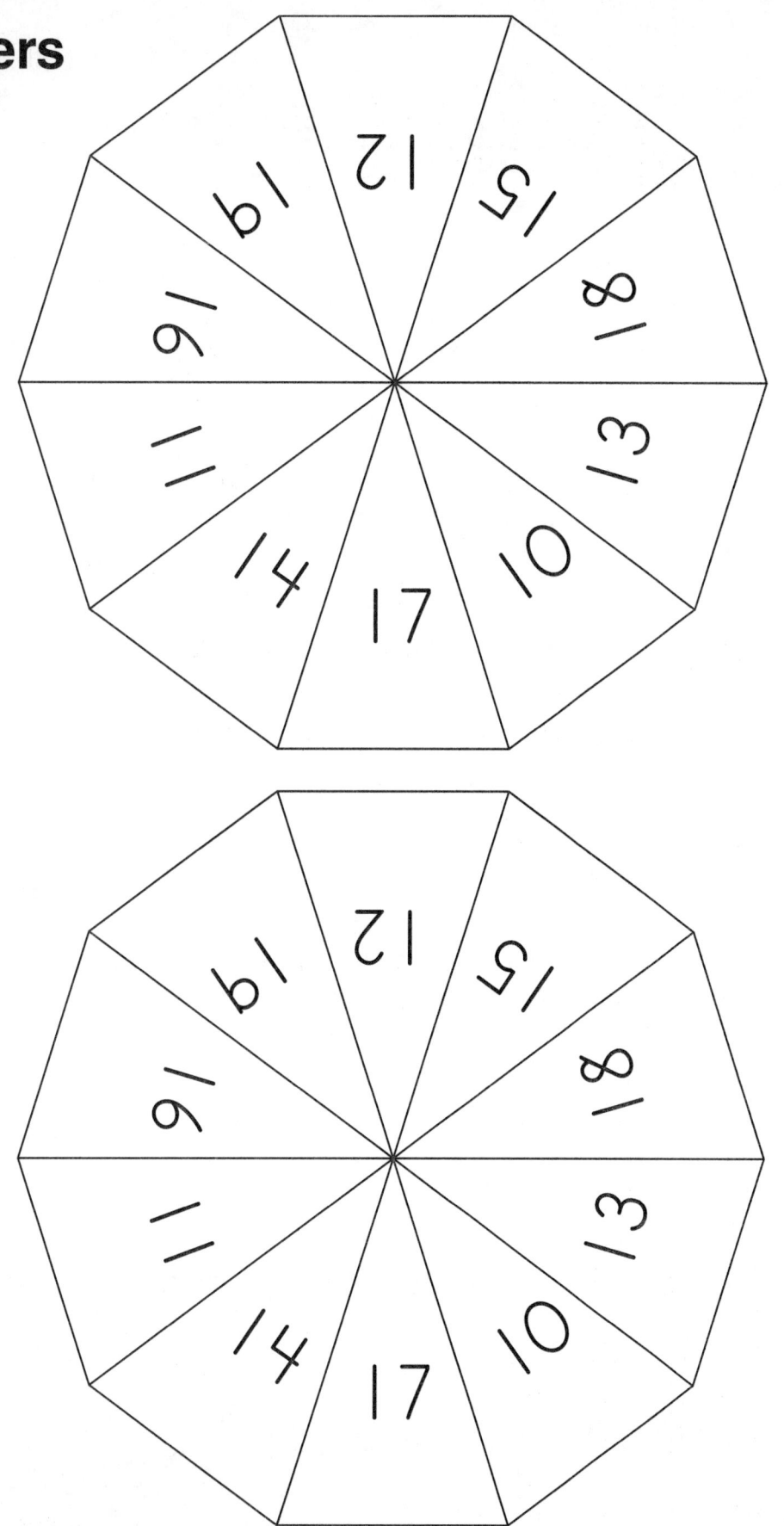

Exploring Addition and Subtraction

Caterpillar Humps

Introducing the Activity

Caterpillars come in many different sizes. Some are long and some are short. We are going to discover all the different ways to construct these caterpillars using up to 20 red and green body parts to attach to a head and a tail. We can record our discoveries using the symbol cards. We will then copy these into our workbooks.

Skills

- ❑ Demonstrate addition and subtraction facts to 20 with objects (A2, S2).
- ❑ Create and solve story problems (A3, S3).
- ❑ Record addition and subtraction activities using symbol/digit cards (A5, S5).
- ❑ Record addition and subtraction activities using written number sentences (A6, S6).
- ❑ Estimate answers to addition and subtraction problems (A7, S7).

How to Make the Activity

- ❑ Copy, laminate, and cut out the four activity cards (pages 35 and 36).
- ❑ Copy the caterpillar heads/tails (page 37). Color and laminate. Cut out as individual pieces.
- ❑ Make 10 copies of the caterpillar humps (page 38). Color half red and half green (or copy onto red and green paper). Laminate and cut out as individual humps. Each student in the group needs up to 20 humps, a caterpillar head, and a tail.
- ❑ Place the equipment into a storage container. Label clearly.

Extra Materials Required

- ❑ two sets of digits 0–9 for each student
- ❑ +, –, = symbol cards
- ❑ 0–9/10–19 spinners
- ❑ Space on the floor or a table (for constructing long caterpillars)
- ❑ Workbooks, pencils for recording

Variations

- ❑ Some students may only want to explore numbers 0–10.
- ❑ This activity can be adapted for written number fact practice to 10 or 20.

#3526 Math in Action ©Teacher Created Resources, Inc.

Caterpillar Humps — Exploring Addition and Subtraction

What a Lot of Humps

- ✔ Take turns to spin the 0–9 spinner.
- ✔ Take that many red humps.
- ✔ Take turns to spin the 0–9 spinner again.
- ✔ Take that many green humps.
- ✔ How many humps altogether on your caterpillar?
- ✔ Record using your small cards.
- ✔ Who has the most humps?

Secret Humps

- ✔ Secretly make a red and green caterpillar.
- ✔ Show it to your partner then hide it.
- ✔ Ask your partner to guess the number combination.
- ✔ Check and record with cards.
- ✔ Exchange roles.

Caterpillar Humps — **Exploring Addition and Subtraction**

Pattern Humps

- ✔ Make a caterpillar with red/green patterns.
- ✔ Guess how many humps of each color you used.
- ✔ Guess how many humps altogether. Check.
- ✔ Record using your small cards.
- ✔ Find another way to make a pattern with these pieces.

Spin Them Away

- ✔ Spin the 10–19 spinner to see how many humps with which to start.
- ✔ Spin the 0–9 spinner to see how many with which to finish.
- ✔ Guess how many humps you have to remove to reach that number.
- ✔ Check and then find a way to record your actions.

Caterpillar Humps | Exploring Addition and Subtraction

Caterpillar Heads and Tails

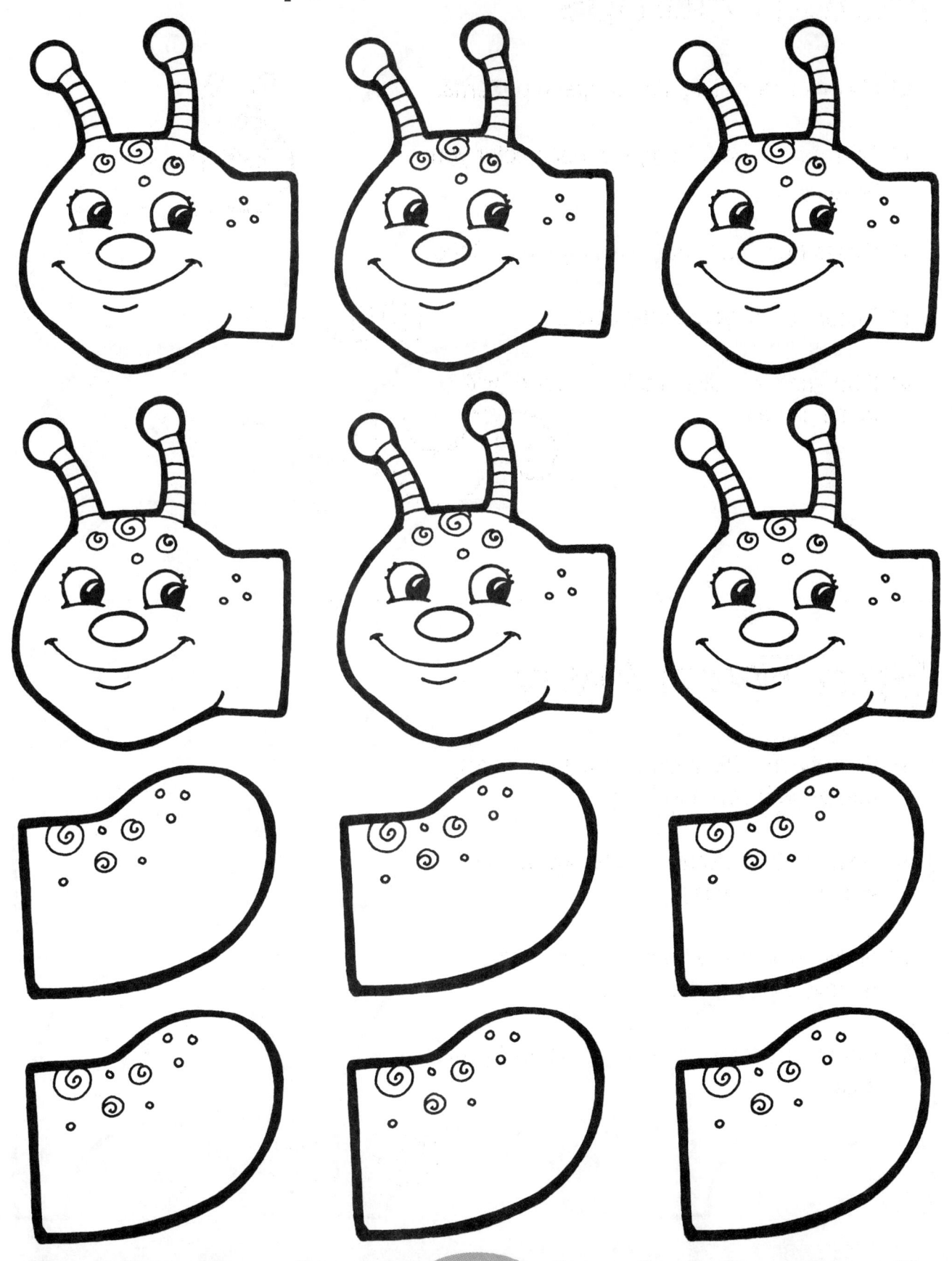

37

©Teacher Created Resources, Inc. | #3526 Math in Action

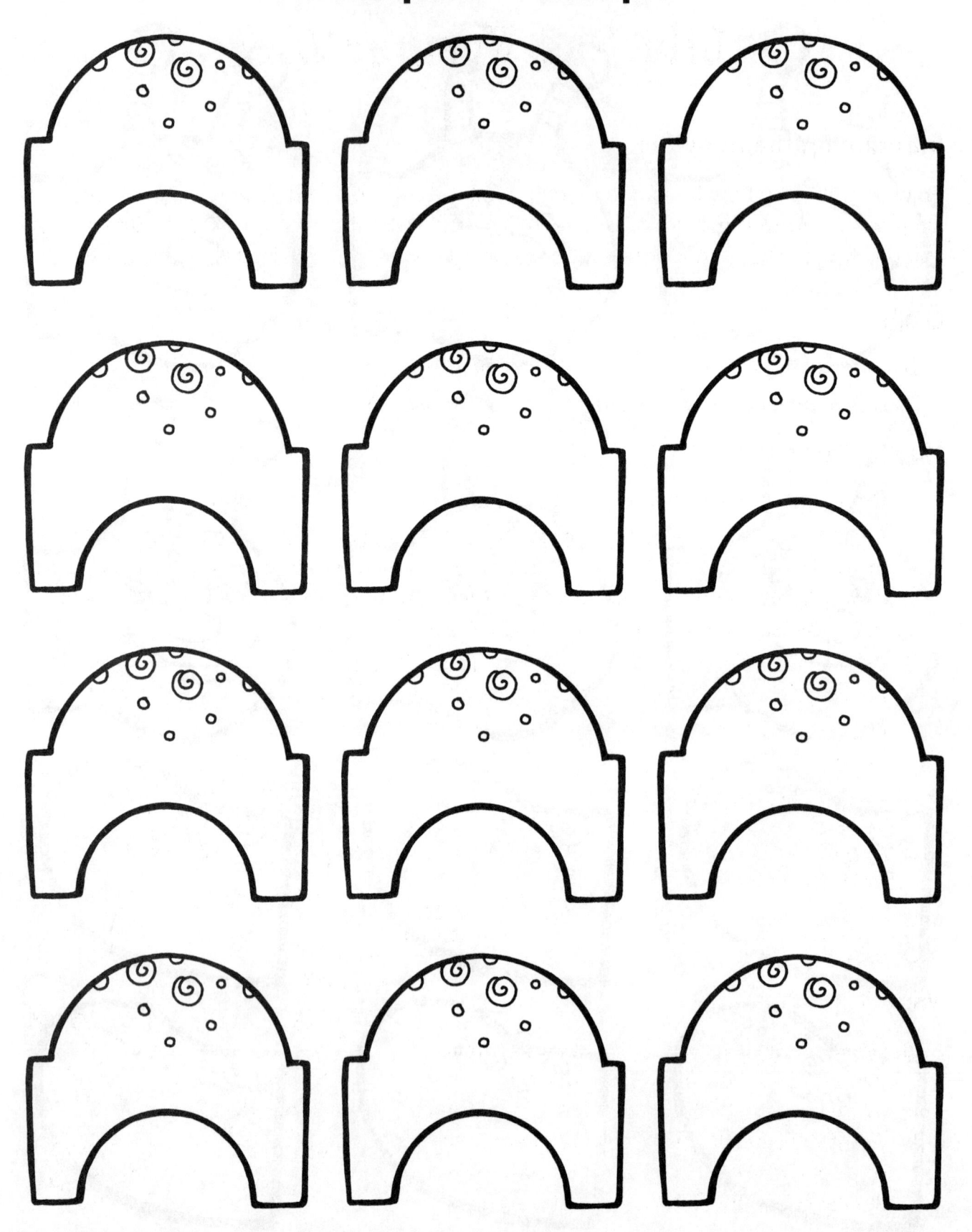

Caterpillar Humps | Exploring Addition and Subtraction

Caterpillar Humps

#3526 Math in Action | 38 | ©Teacher Created Resources, Inc.

Exploring Addition and Subtraction

Lily Pad Number Line

Introducing the Activities

Frogs love to jump from lily pad to lily pad. Pretend you are a frog. Where will you start jumping? Where will you finish? The numbers on these lily pads can help us count forward and backward. They can also help us when we are adding up or taking numbers away.

Skills

- Use a number line to solve addition problems (A8).
- Use a number line to solve subtraction problems (S8).

How to Make the Activity

- Copy, laminate, and cut out the four activity cards (pages 40 and 41).
- Each student will need a copy of the lily pads (page 42). Four strips pasted together make a 0–20 number line. The fifth strip can be used later to make a longer line. Color, then cut along the dotted lines. Paste together to make one long line of 20 lily pads. Paste the Start Frog to the left of the first lily pad.
- Write the numbers from 1 onward on the lily pads to the right. These can now be used for counting activities, then folded up and placed under desks or inside a tray or workbook for later use.

Extra Materials Required

- a small plastic 3-D frog counter (or any small counter) for each student
- dice and 0–9/10–19 spinners
- calculators

Variation

- Construct lily pad number lines to 30 or more, depending on how much desk space is available. One 0–20 number line is just under 3 feet (90 cm) long. One 0–30 number line will be about 4 feet (130 cm). Alternatively, reduce the page by 70% to make a smaller version.

Lily Pad Number Line   **Exploring Addition and Subtraction**

Lily Pad Leap

- ✔ Put your counter on Start.
- ✔ Throw a die and jump along lily pads to match.
- ✔ Throw the die again.
- ✔ Guess which number you will land on when you jump that many more lily pads.
- ✔ Check by jumping and saying the number story aloud.

    e.g., six and three make nine

- ✔ Check again with a calculator.

Jump Back

- ✔ Spin the 10–19 spinner.
- ✔ Start at that lily pad.
- ✔ Spin the 0–9 spinner.
- ✔ Guess where you will land when you jump back by that many lily pads.
- ✔ Check by jumping, saying the number story aloud.

    e.g., seventeen take away four makes thirteen

- ✔ Check again with a calculator.

#3526 Math in Action   ©Teacher Created Resources, Inc.

Lily Pad Number Line   **Exploring Addition and Subtraction**

Jump to 20

- ✔ Work with a partner.

- ✔ Take turns to throw a die and jump along that many lily pads, saying your number **actions aloud.**

- ✔ Keep playing until someone jumps to 20.

    e.g., nine and four more makes thirteen

- ✔ Play again, but this time start at 20 and jump back to 0.

More Jumps

- ✔ Work with a partner.

- ✔ Take turns to throw the die and jump that many times.

- ✔ How many more jumps did the furthest frog jump?

- ✔ Guess first and then check.

- ✔ Find a way to check this difference with the calculator, too.

- ✔ Continue playing until one frog jumps to 20.

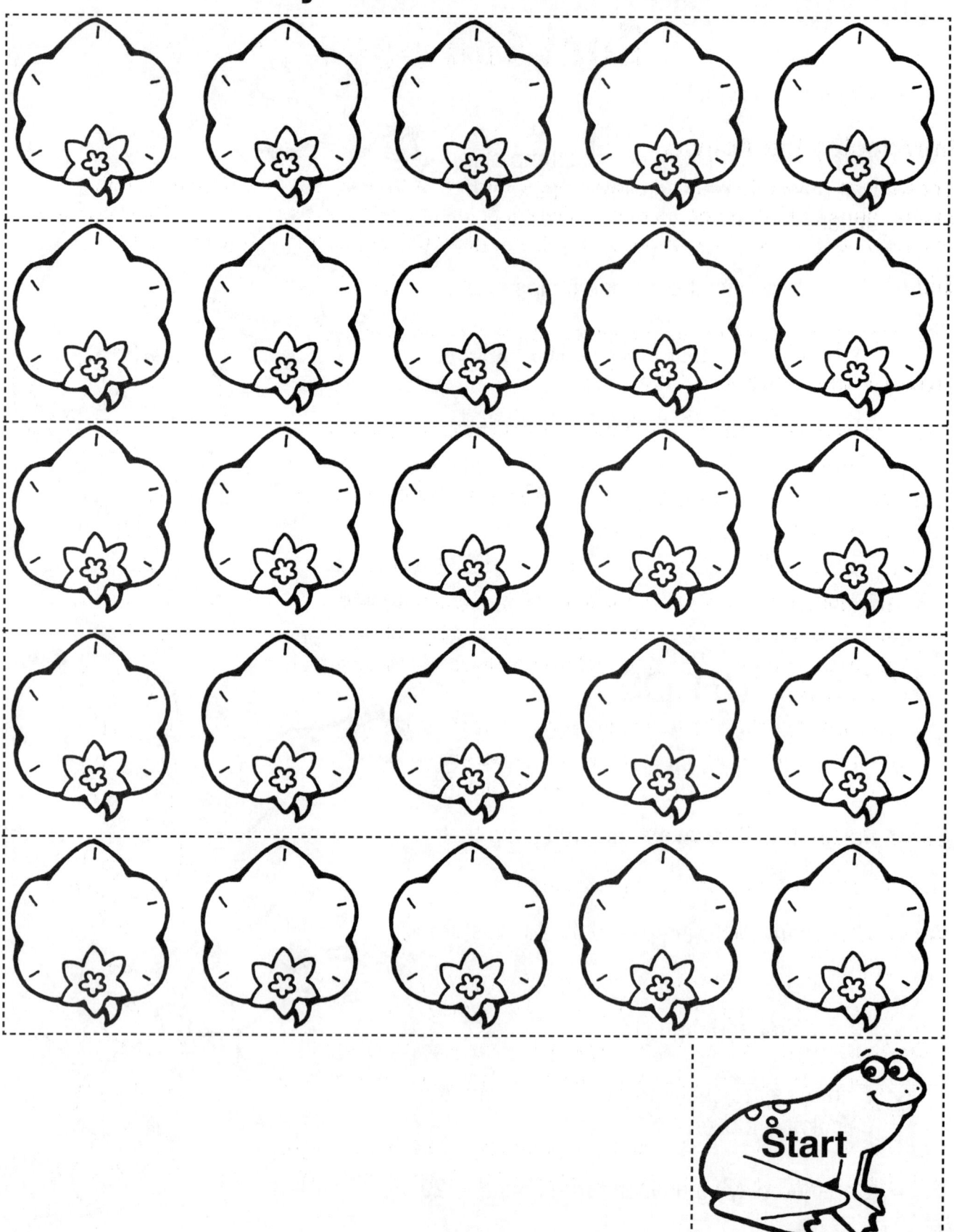

Lily Pad Number Line

Exploring Addition and Subtraction

Lily Pad Number Line

Exploring Addition and Subtraction

Sort the Eggs

Introducing the Activity

These hens have mixed up all of their eggs. How can we sort and match them with the mother hens? (Encourage as many different solution strategies as possible.) Can you think of the number names in your head, or will you need objects to help you work it out?

Skill

- Recall addition facts to 10 using a range of strategies (A9).

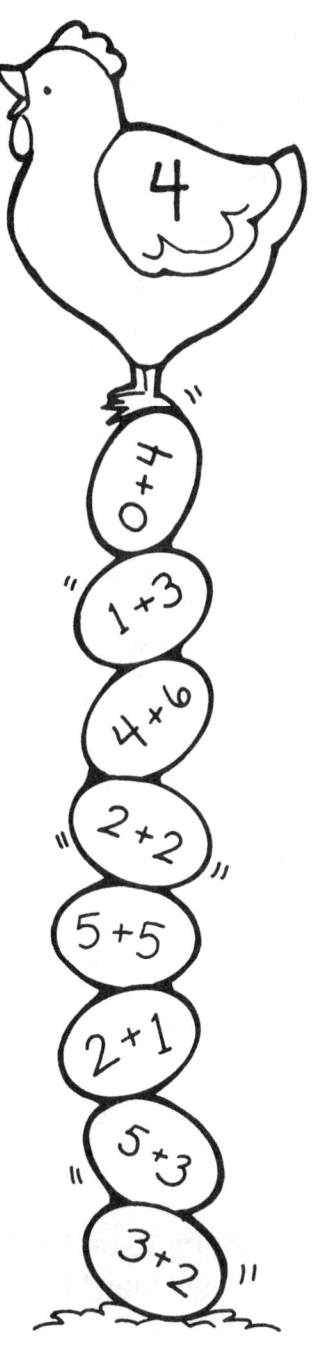

How to Make the Activity

- Copy, laminate, and cut out the four activity cards (pages 44 and 45).
- Make 11 copies of the large hen (page 46). Color and label each one from 0–10. Laminate.
- Make five copies of the eggs (page 47) on yellow cardboard.
- You need a list of addition facts to write on each egg. Ask students to discover these for you. For example, place 0–10 counters into two groups or patterns to discover addition facts from 0–10 (0 + 1, 3 + 2, 5 + 3, 4 + 6, 5 + 5, . . .). As you discover a fact, write it onto an egg. Once of all the eggs have number facts, laminate and then cut out as individual eggs. Store the eggs together in a labeled container with a lid.

Extra Materials Required

- 10 counters for discovering number facts

Variations

- Subtraction facts to 10 can be explored by copying more eggs (red cardboard) and recording subtraction facts (S9).
- Addition facts to 20 can be explored by copying more eggs (green cardboard) and hens (numbered 11–20) and recording addition facts to 20 (A10).
- Subtraction facts to 20 can be explored by copying more eggs (blue cardboard) and recording subtraction facts to 20 (S10).

Sort the Eggs   **Exploring Addition and Subtraction**

Sort the Eggs

- ✔ Mix all the eggs face up.

- ✔ Sort them with the matching mother hens.

- ✔ Does each hen have the same number of eggs?

- ✔ Why?

Scrambled Eggs

- ✔ Mix up the eggs face down.

- ✔ Take two hen cards each.

- ✔ Take turns to turn over two eggs.

- ✔ Keep any eggs that belong to your hens.

- ✔ Turn the eggs over again if they do not match a hen.

- ✔ Try to be the first person to have six matching eggs.

Sort the Eggs   **Exploring Addition and Subtraction**

Large Eggs

- ✔ Mix up the eggs face down.
- ✔ Turn over an egg each.
- ✔ Who has the smallest number?
- ✔ Who has the largest number?
- ✔ Does anyone have the same number?
- ✔ Repeat at least five times.

Name the Eggs

- ✔ Take one hen card and sit with your back to the group.
- ✔ Now call out number names for your hen.
- ✔ Ask the rest of your group to give you any eggs you correctly name.
- ✔ Try to collect all the matching eggs.
- ✔ Exchange roles.

Sort the Eggs **Exploring Addition and Subtraction**

#3526 Math in Action ©Teacher Created Resources, Inc.

Sort the Eggs Exploring Addition and Subtraction

Exploring Addition and Subtraction

Fish Bits

Introducing the Activity

These fish have all been broken into eight pieces. Can you put them back together by finding the matching number facts? Six body pieces match the number on the tail of each fish.

Skills

- ❑ Recall addition facts to 10 using a range of strategies (A9).
- ❑ Recall subtraction facts to 10 using a range of strategies (S9).

How to Make the Activity

- ❑ Make six copies of the fish (page 49) onto blue cardstock paper.
- ❑ Write a number from 0–10 on the fish's tail. Then fill in the fish's body with the matching number facts (do not write in the head). Use some of the number facts below:

0:	0 + 0	0 – 0	1 – 1	2 – 2	3 – 3	4 – 4	5 – 5	6 – 6	7 – 7	8 – 8	9 – 9	10 – 10
1:	0 + 1	1 + 0	1 – 0	2 – 1	3 – 2	4 – 3	5 – 4	6 – 5	7 – 6	8 – 7	9 – 8	10 – 9
2:	0 + 2	2 + 0	1 + 1	2 – 0	3 – 1	4 – 2	5 – 3	6 – 4	7 – 5	8 – 6	9 – 7	10 – 8
3:	0 + 3	3 + 0	1 + 2	2 – 1	3 – 0	4 – 1	5 – 2	6 – 3	7 – 4	8 – 5	9 – 6	10 – 7
4:	0 + 4	4 + 0	1 + 3	3 – 1	2 + 2	4 – 0	5 – 1	6 – 2	7 – 3	8 – 4	9 – 5	10 – 6
5:	0 + 5	5 + 0	1 + 4	4 – 1	2 + 3	3 + 2	5 – 0	6 – 1	7 – 2	8 – 3	9 – 4	10 – 5
6:	0 + 6	6 + 0	1 + 5	5 – 1	2 + 4	4 + 2	3 – 3	6 – 0	7 – 1	8 – 2	9 – 3	10 – 4
7:	0 + 7	7 + 0	1 + 6	6 – 1	2 + 5	5 + 2	3 – 4	4 + 3	7 – 0	8 – 1	9 – 2	10 – 3
8:	0 + 8	8 + 0	1 + 7	7 – 1	2 + 6	6 + 2	3 – 5	5 + 3	4 + 4	8 – 0	9 – 1	10 – 2
9:	0 + 9	9 + 0	1 + 8	8 – 1	2 + 7	7 + 2	3 – 6	6 + 3	4 + 5	5 – 4	9 – 0	10 – 1
10:	0 + 10	10 + 0	1 + 9	9 – 1	2 + 8	8 + 2	3 + 7	7 + 3	4 + 6	6 – 4	5 + 5	10 – 0

- ❑ Write the family number (0–10) on each fish tail. Laminate and cut each fish into eight pieces. Then store in a container and label clearly.

Extra Materials Required

- ❑ none

Variations

- ❑ Copy more fish (red cardstock paper) and record random +/– facts to 20 (A10, S10).
- ❑ Copy more fish (yellow cardstock paper). Write random numbers from 0–10 on the six bits, excluding the head and tail. Do not cut this fish into pieces. Leave it as a complete fish. Use the fish pieces from the earlier game. Find number name pieces to match

#3526 Math in Action

©Teacher Created Resources, Inc.

Fish Bits   **Exploring Addition and Subtraction**

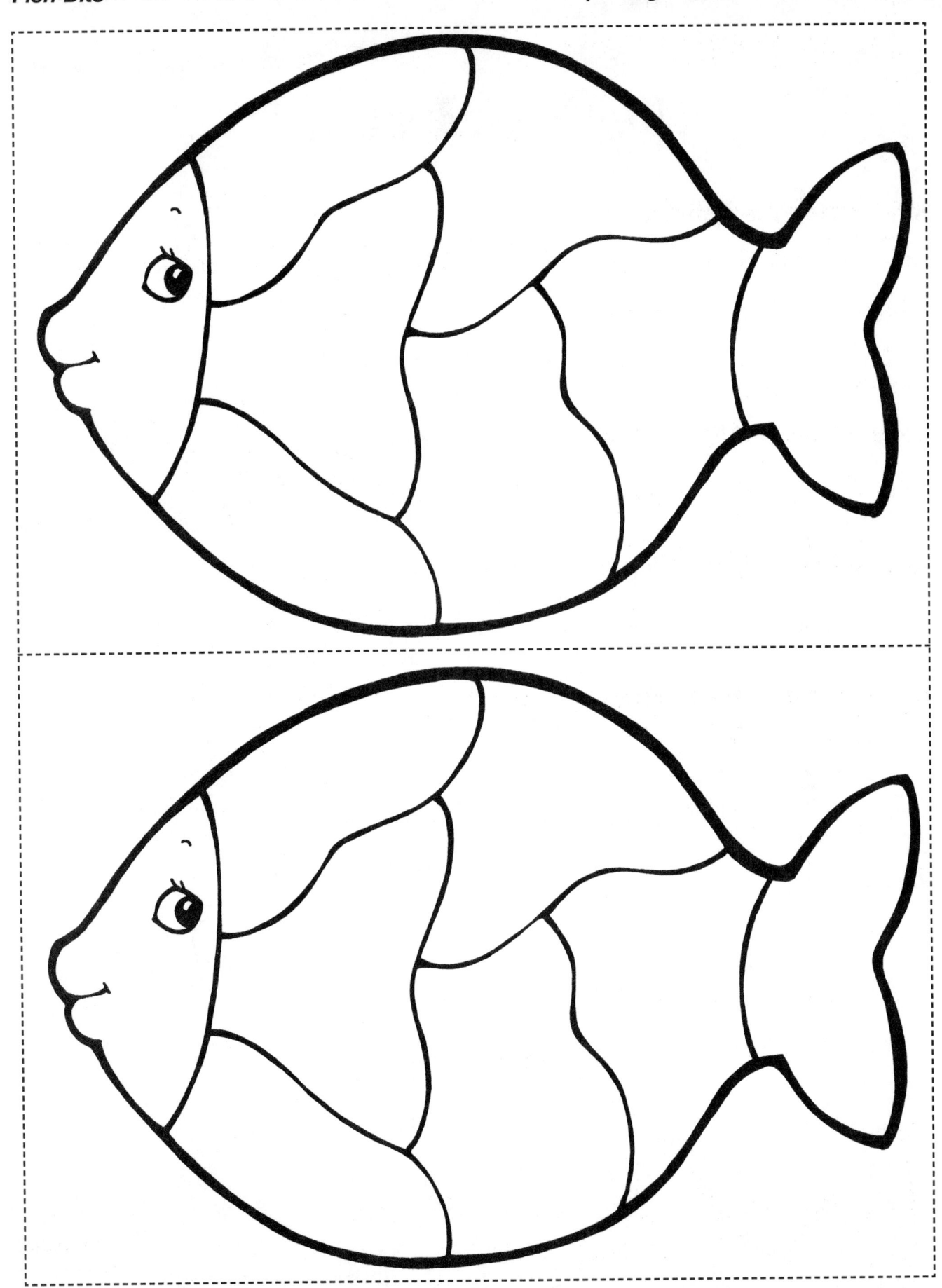

©*Teacher Created Resources, Inc.*   49   *#3526 Math in Action*

Exploring Addition and Subtraction

Digit Card Games

Introducing the Activities

Once you think you can remember number facts without always using objects to help you work out answers, try some of the following games using just the digit cards 0–9. All four games start with two sets of digit cards face down in the middle of your group.

Skills

- Recall addition facts to 20 using a range of strategies (A10).
- Recall subtraction facts to 20 using a range of strategies (S10).

How to Make the Activities

- Copy the four activity cards (pages 51 and 52). Laminate and cut out.

Extra Materials Required

- two sets of digit cards 0–9 for each group (from previous activities—see page 14)
- a pile of building bricks (e.g., multilink) for "Add and Build" on page 50

Variations

- Add and build: When the group model is built, estimate the total number of pieces used. Check by counting. Take three or four cards. Add and build (A11).
- Smallest: Add the two numbers. The smallest number wins (A10).
- The 15 Game: Find three cards that add to 16, 18, . . . (A10). Find four cards that add to 20, 17, . . . (A11).
- Count down: Start at a different number (e.g., 16) (S11).
- Make up your own rules for a game.

#3526 *Math in Action* ©Teacher Created Resources, Inc.

Digit Card Games — Exploring Addition and Subtraction

Add and Build

- ✔ Put a pile of blocks in the middle.
- ✔ Turn over two digit cards each.
- ✔ Add them together and take that many blocks.
- ✔ Build one large model together.
- ✔ Guess how many blocks.
- ✔ Check your guesses.

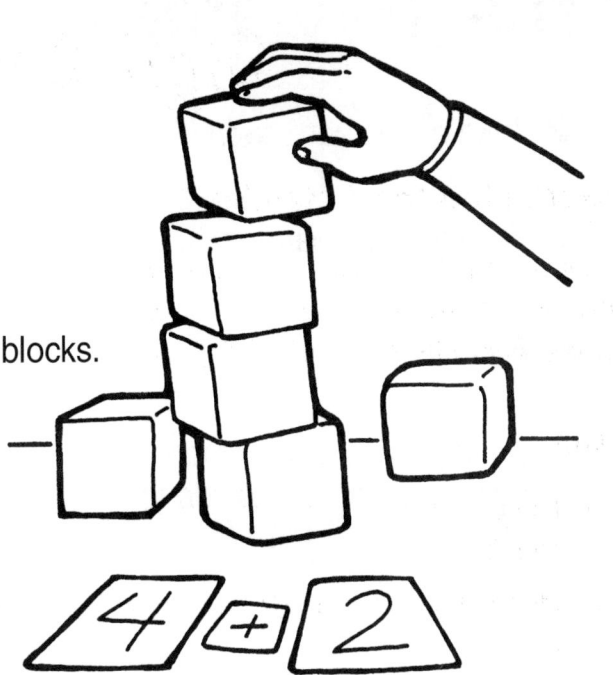

Smallest

- ✔ Take two cards each.
- ✔ Find the difference.
- ✔ The person with the smallest difference keeps all the cards used.
- ✔ Keep playing until there are no cards left.

Digit Card Games

Exploring Addition and Subtraction

The 15 Game

- ✔ Take a digit card each.

- ✔ Take another card each.

- ✔ Add it to your first card.

- ✔ Keep playing until someone has three cards that add to 15.

Count Down

- ✔ Start at 20.

- ✔ Take turns to turn a card over and subtract.

- ✔ See who gets to zero first.

Mental Recall Ideas

Dominoes

- Play with a partner. Turn over a domino each. Who has the larger number of spots? The least spots? Tell each other about your spots using number sentences. (e.g., Five spots and three spots equal eight spots altogether. That is more than four and two spots.) (A9)

- Reveal a domino to a partner for a short time (e.g., two seconds). Ask your partner to tell you what he or she saw. (e.g., I saw ten spots–six and four.) (A9)

- Race against a three-minute timer. Work with a partner. Sort all the dominoes into number name groups. (e.g., All the number names for eight go here.) (A9)

- Turn all the dominoes face down. Play a memory game. Take turns to turn over two dominoes. If they name the same number using addition and/or subtraction, you keep them. If not, turn them face down again. The winner is the person with the most dominoes at the end of the game. (A9, S9)

Dice

- Play with a partner. Throw two dice. Add the numbers shown. Whoever has the larger (or smaller) total wins a counter. (A9)

- Play with a partner. Throw two dice. Subtract the smaller number from the larger number shown. Whoever has the larger (or smaller) difference wins a counter. (S9)

Spinners

- Play with a partner. Take turns to spin the 10–19 spinner then the 0–9 spinner. Take the smaller number shown from the larger number. Whoever has the smaller (or larger) difference wins a counter. (S9)

Tortoise Bingo

Each student needs a copy of the Bingo Tortoise (page 54) and six small counters. The caller needs a copy of the Caller's Cards (pages 55 and 56), cut up, and placed in a small envelope.

- Write random numbers 0–10 in the six spaces on your tortoise's back. Cover each number with a counter when you hear a number name that matches. The first person to cover all six numbers calls out "Tortoise Bingo." (A9, S9)

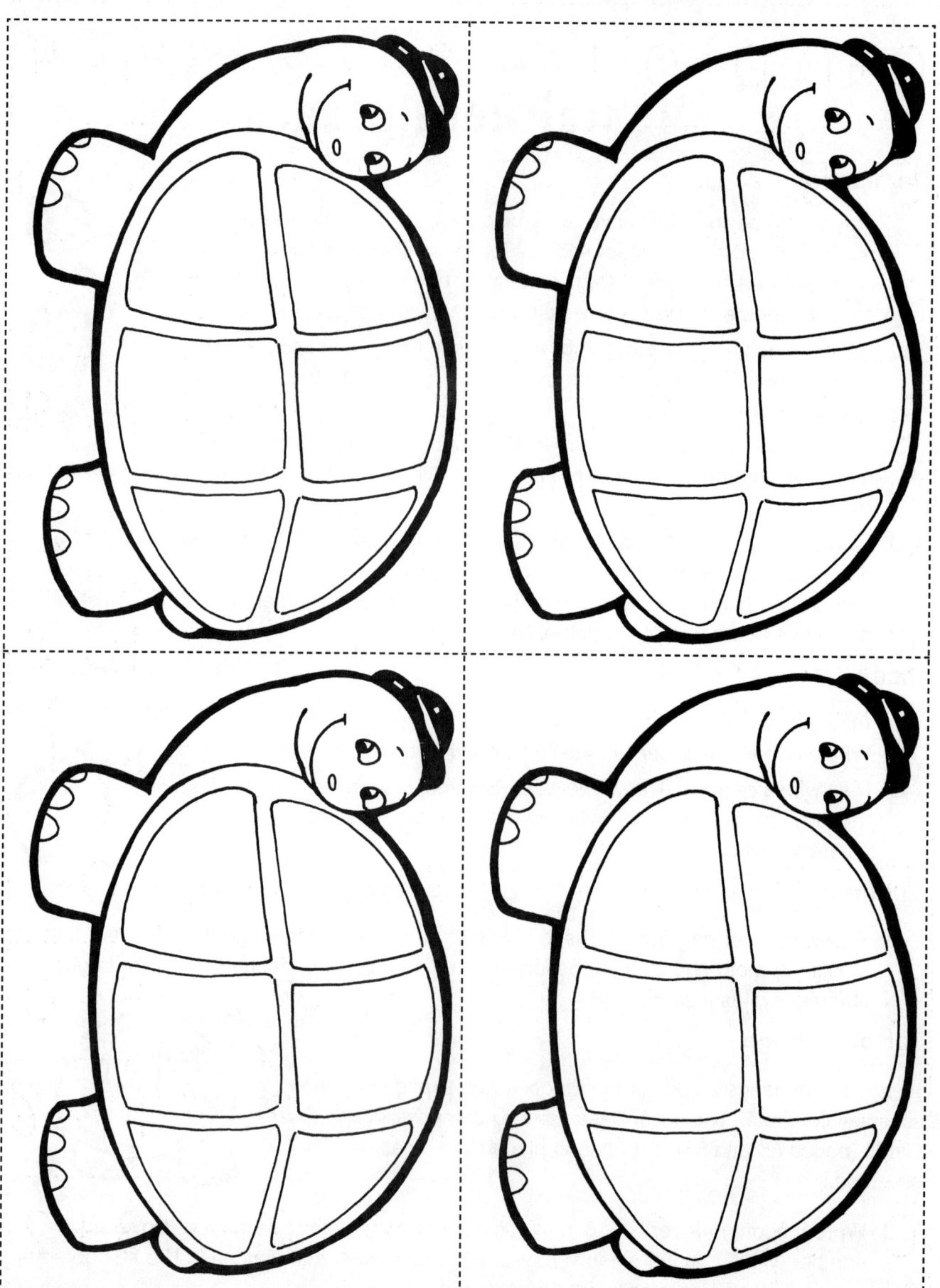

Mental Recall Ideas   **Exploring Addition and Subtraction**

Mental Recall Ideas   **Exploring Addition and Subtraction**

Mental Recall Ideas — Exploring Addition and Subtraction

0 + 0	0 − 0	1 − 1	2 − 2	3 − 3	4 − 4
5 − 5	6 − 6	7 − 7	8 − 8	9 − 9	10 − 10
0 + 1	1 + 0	1 − 0	2 − 1	3 − 2	4 − 3
5 − 4	6 − 5	7 − 6	8 − 7	9 − 8	10 − 9
0 + 2	2 + 0	1 + 1	2 − 0	3 − 1	4 − 2
5 − 3	6 − 4	7 − 5	8 − 6	9 − 7	10 − 8
0 + 3	3 + 0	1 + 2	2 + 1	3 − 0	4 − 1
5 − 2	6 − 3	7 − 4	8 − 5	9 − 6	10 − 7
0 + 4	4 + 0	1 + 3	3 + 1	2 + 2	4 − 0
5 − 1	6 − 2	7 − 3	8 − 4	9 − 5	10 − 6
0 + 5	5 + 0	1 + 4	4 + 1	2 + 3	3 + 2

Mental Recall Ideas **Exploring Addition and Subtraction**

5 – 0	6 – 1	7 – 2	8 – 3	9 – 4	10 – 5
0 + 6	6 + 0	1 + 5	5 + 1	2 + 4	4 + 2
3 + 3	6 – 0	7 – 1	8 – 2	9 – 3	10 – 4
0 + 7	7 + 0	1 + 6	6 + 1	2 + 5	5 + 2
3 + 4	4 + 3	7 – 0	8 – 1	9 – 2	10 – 3
0 + 0	8 + 0	1 + 7	7 + 1	2 + 6	6 + 2
3 + 5	5 + 3	4 + 4	8 – 0	9 – 1	10 – 2
0 + 9	9 + 0	1 + 8	8 + 1	2 + 7	7 + 2
3 + 6	6 + 3	4 + 5	5 + 4	9 – 0	10 – 1
0 + 10	10 + 0	1 + 9	9 + 1	2 + 8	8 + 2
3 + 7	7 + 3	4 + 6	6 + 4	5 + 5	10 – 0

Addition and Subtraction Check-Up

Resource
copies of page 58

Activity
Use this check-up with small groups or with the whole class. Record student responses on the Skills Record Sheet (page 95). The following are suggested instructions and sample teacher's comments when developing student profiles.

"Look at the fish bowls. Draw four tadpoles in the first one and five tadpoles in the second one. Write the numbers that tell you how many there are, in the spaces, as a number sentence." (A4)

Sample Profile Comments
Records addition activities up to 10 using words and digits.

"Look at the domino. Draw eight spots altogether. How many spots on the left? How many spots on the right? Fill in the missing numbers to make a number story that matches the domino." (S4)

Sample Profile Comments
Records subtraction activities to 10 using words and digits.

"Look at the apple trees. What is missing? *(apples)* Draw in 13 apples on the first tree."

"Draw four apples on the second tree. Underneath write a number sentence to match." (A2,5)

Sample Profile Comments
Demonstrates and records addition actions up to 20 and records using symbols and digits.

"Find the long snake. How many spotted parts are there? *(19)* Cross out some spotty parts to show nineteen take away five. Write the number sentence to match underneath." (S2,5)

Sample Profile Comments
Demonstrates subtraction actions to 20. Records using symbols and digits.

"At the bottom of your page, there are two hens. Write a number between 0–20 on each hen. Write your own number names on the eggs to match each mother hen." (A9,10; S9,10)

Sample Profile Comments
Recalls addition and subtraction facts to 10 and records these using written number names.

Recall of number facts to 20 is still difficult, but he knows how to use concrete materials to help solve these problems.

"Turn your page over. In the top half, make up your own picture story about adding any numbers you like, even numbers past 20. Record your story with a number sentence." (A3,6)

"In the bottom half, make up your own picture story about taking away any numbers you like, even numbers past 20. Record your story with a number sentence." (S3,6)

Sample Profile Comments
Joe really enjoys exploring numbers past 20. He can create and record addition and subtraction actions using his own strategies and recording system.

Addition and Subtraction Check-Up

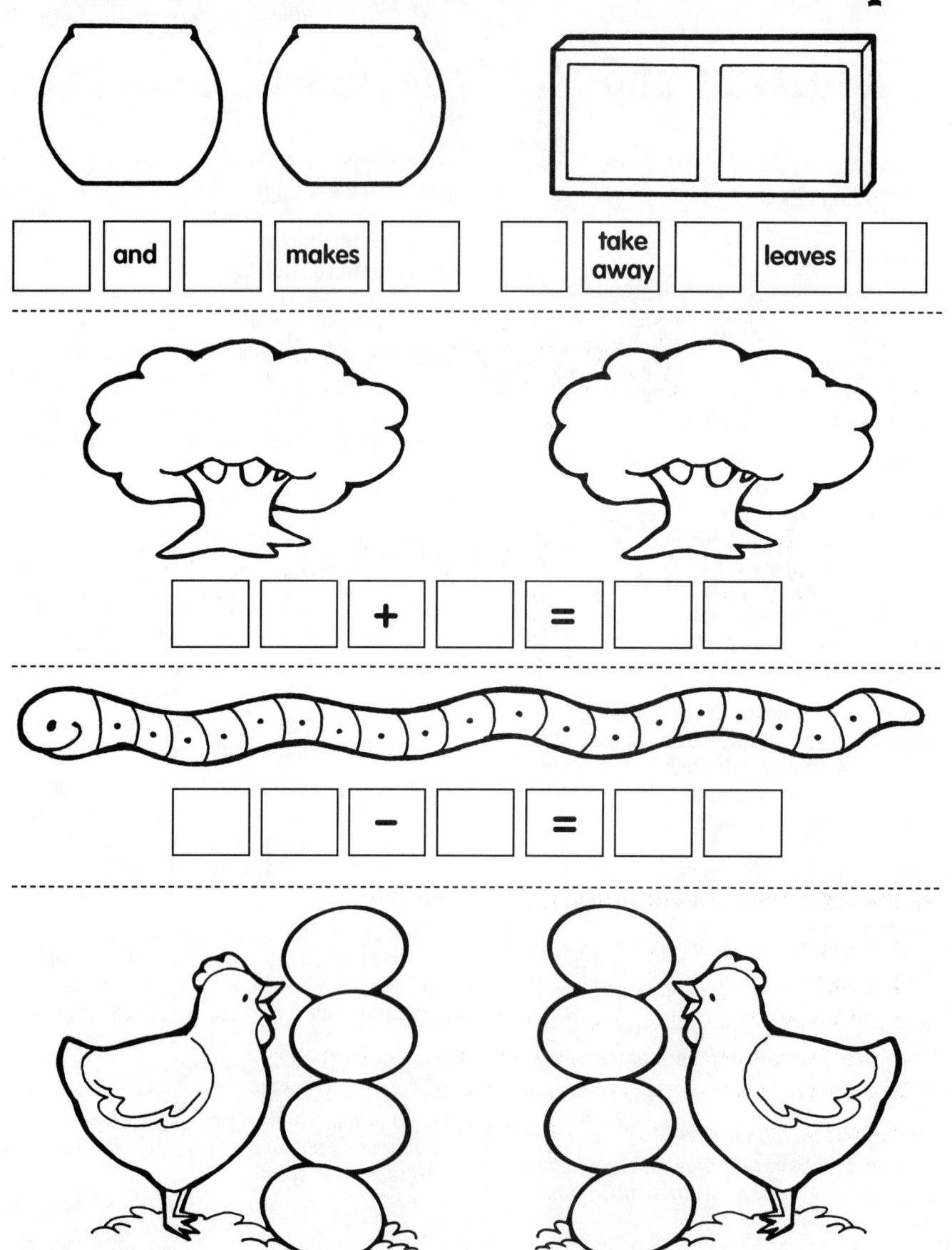

☐ and ☐ makes ☐   ☐ take away ☐ leaves ☐

☐☐ + ☐ = ☐☐

☐☐ − ☐ = ☐☐

Exploring Multiplication

In this unit, your students will do the following:

- ❑ Reinforce counting by 2s, 5s, and 10s.
- ❑ Model multiplication story problems as equal groups or rows.
- ❑ Understand that multiplication is just repeated addition.
- ❑ Discover multiplication combinations using informal language.
- ❑ Estimate answers to multiplication story problems.
- ❑ Record in three stages using informal word cards, formal word cards, and written number sentences.

Exploring Multiplication

Multiplication "Active-ities"

Counting Patterns [CLASS] M1

Activity

Review counting aloud by 1s (e.g., students), 2s (e.g., pairs of eyes), 5s (e.g., fingers on one hand), and 10s (e.g., toes on one person) forward and backward as far as you can go. Try counting by 3s or 4s, too. Can anyone count by 6s?

Bunch Up [CLASS] M2,3

Resources

tape recorded music (optional)

Activity

Discuss what "bunch up" means—forming a group with that many people in it. The whole class then moves around, preferably outside, to the music. When the music stops, call out a number between 0 and 10. Everyone races to bunch up.

Any extra students are given the job of checking the numbers within each group, counting how many groups and counting how many students altogether. They call out their findings to the whole group. (e.g., 4, 8, 12, 16, 20, 24) There are six groups of four. That is 24 altogether. Repeat using other numbers.

Rows [CLASS] M2,3

Resources

tape recorded music (optional)

Activity

Again, this is best done outside or in a large hall. Discuss what a row is and in which direction the rows will face. When the music stops, call out a number between 0 and 10. Everyone races to form straight rows containing that many people.

Any extra students count how many in each row, how many rows altogether, and how many students altogether. They call out their findings to the whole group. (e.g., 5 and 5 and 5 and 5 and 5.) There are five rows of five. That makes 25 altogether. Repeat using other numbers.

Exercises [CLASS] M2,3

Activity

Stand in a large space not too near to each other. Call out an exercise (e.g., touch your feet, stretches, etc.). Next, ask someone to call out a number from 0–10. This is how many of each exercise you will do. Call out another number from 0–10. This tells you how many times you will repeat your exercise (e.g., six stretches three times).

Ask a student to count the total number of exercises you do altogether. Repeat using other numbers.

Multiplication "Active-ities"

Flowers   GROUP   M2,3,4,8

Resources
10 plastic yogurt containers, 60 craft sticks, craft paper to make petals and leaves, scissors, glue or paste for each group

Activity
Each group decorates the ends of their sticks with cut-out paper petals and leaves. When these are dry, use them to form vases of flowers with the same number of flowers in each vase. Take turns to create a multiplication number story and ask the others to model this with flowers. (e.g., My grandma has four vases with seven flowers in each vase.) Everyone then guesses how many flowers there are altogether. Check by counting. (e.g., There are 7 and 7 and 7 and 7 flowers. Grandma has 28 flowers altogether.)

Towers   GROUP   M1,2,3,4,8

Resources
colored building blocks (e.g., multilink blocks)

Activity
Race each other to build tall towers where every group of five blocks is a different color. Estimate how many pieces you used altogether. Count and check. Try counting by 5s. Describe your tower using grouping terms. (e.g., My tower has five red, five yellow, five blue, and five more yellow blocks. That is four groups of five. I used 20 blocks altogether.)

Variation
Build long caterpillars or snakes where every group of 2, 3, 4, 5, and 10 blocks is another color. What is the largest number of blocks you can use? Did each group have exactly the same number of blocks in them?

Previous Resources   GROUP   M1,2,3,4,8

Resources
Use the addition and subtraction resources to model equal groups, rows or columns (e.g., This caterpillar has 18 humps altogether. There are 6 red, 6 green and 6 more red humps). (See pages 36 and 38.)

Number Roll-Ups   PAIR   M1

Resources
Number Roll-Ups (page 62), cut into lengths along the dashed lines (*Note:* Glue the ends together to make one long strip. Fold along the lines starting from the bottom.)

Activity
Practice counting aloud by 10s (fingers), unrolling the strips as you count.

How fast can you go? Can you count backward too?

Variation
Make your own number roll-ups for practicing counting by 2s (e.g., eyes) or 5s (e.g., pentagon sides).

Multiplication Activities *Exploring Multiplication*

Number Roll-Ups
x10

Paste bottom of first strip here.

Paste top of third strip here.

Exploring Multiplication

Dog Bones

Introducing the Activity

These dogs love to play with up to five bones each. They like to have exactly the same amount so they do not squabble. We will discover how many bones altogether if the dogs have two bones each, three each, and so on. We will make up stories about them and practice recording our discoveries using cards for *groups of* and *makes*. Who can think of a story about dogs/bones to start? *(Discuss and demonstrate.)*

Skills

- ❏ Model multiplication as equal groups (M2).
- ❏ Use language *groups of* and *makes* (M3).
- ❏ Create, solve multiplication story problems (M4).
- ❏ Record multiplication activities using informal word/digit cards (M5).
- ❏ Estimate answers to simple multiplication problems (M8).

How to Make the Activity

- ❏ Copy, laminate, and cut out the four activity cards *(pages 64 and 65)*.
- ❏ Make five copies of the dogs (page 66) so you have 10 dogs altogether. Color, laminate, and cut out as cards. These will be ideal for two students to use, but can be used by up to five students in a group.
- ❏ Make five copies of the bones (page 67). Color, laminate, and cut out as 50 individual bones.
- ❏ Copy the *groups of/makes* cards (page 68). Laminate and cut out along the dashed lines.
- ❏ Place the dogs/bones in a storage container and label clearly. Place the word cards into a separate container.

Extra Materials Required

- ❏ two sets of the digits 0–9 from previous activities (page 14) for each student
- ❏ 0–9 spinner (page 15) for each pair

Variations

- ❏ Make up your own dog and bone stories about equal groups.
- ❏ Use 50 large round counters and pretend these are balls with which the dogs are to play, in place of the bones.
- ❏ Copy more dogs and bones to explore multiplication facts beyond 50.

©Teacher Created Resources, Inc. #3526 Math in Action

Dog Bones   Exploring Multiplication

Match Those Bones!

- ✔ Spin the spinner.
- ✔ Take that many dogs.
- ✔ Give each dog two bones.
- ✔ How many groups of bones?
- ✔ How many bones in each group?
- ✔ How many bones altogether?
- ✔ Check and then find a way to record your actions using the small cards.

How Many Altogether?

- ✔ Make the first part of a number story with cards.

    e.g., | 4 | groups of | 3 |

- ✔ Tell your dog story to a friend.

    e.g.,   Four dogs had a birthday.

- ✔ Their owner gave them three bones each.

- ✔ Exchange cards. Race to model the new story with dogs and bones.

- ✔ Guess how many bones altogether. Check. Record the total number.

    e.g., | 4 | groups of | 3 | makes | 1 | 2 |

64

#3526 Math in Action   ©Teacher Created Resources, Inc.

Dog Bones — Exploring Multiplication

Hidden Bones

✔ Give some dogs an equal number of bones.

✔ Make a number sentence to match.

e.g., | 4 | groups of | 2 | makes | 8 |

✔ Turn over the answer card.

e.g., | 4 | groups of | 2 | makes | |

✔ Hide the bones behind each dog. Ask a friend to look at your cards and try to compute how many bones altogether.

✔ Check and then exchange roles.

How Many Dogs?

✔ My dogs are very happy.

✔ They have 20 bones altogether.

✔ They each have the same number of bones.

✔ How many dogs could there be?

✔ Repeat using other numbers.

©Teacher Created Resources, Inc. — #3526 Math in Action

Dog Bones Exploring Multiplication

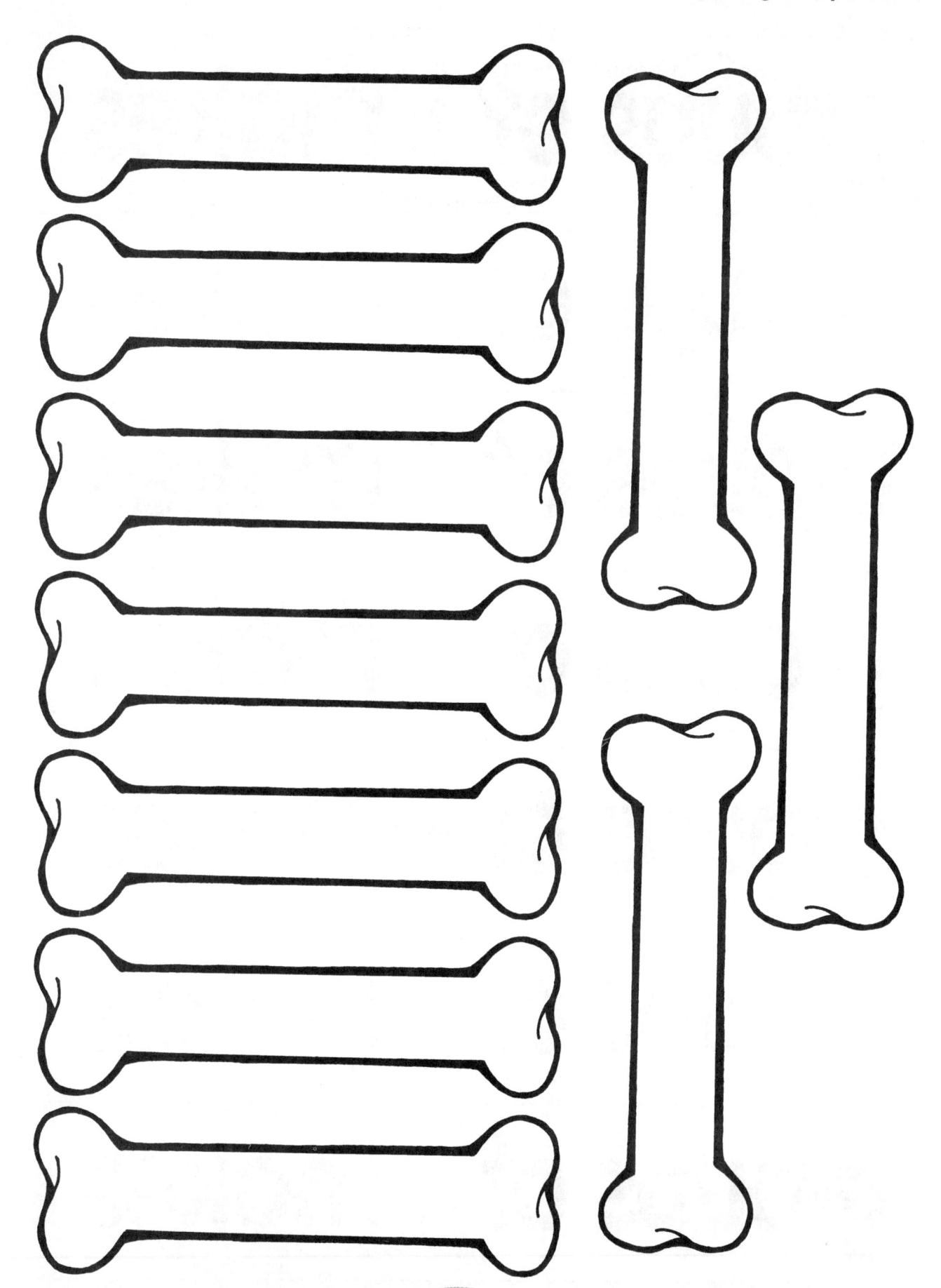

Dog Bones

Exploring Multiplication

Dog Bones   **Exploring Multiplication**

groups of	makes
groups of	makes
groups of	makes
groups of	makes
groups of	makes
groups of	makes
groups of	makes

Exploring Multiplication

Caterpillars

Introducing the Activity

These caterpillars are missing their rows of legs. Pretend the pegs are their legs. We will discover how many legs they have altogether when they each have two legs or more. We will make up stories about them and practice recording our discoveries using *rows of* and *makes* cards. Close your eyes and think of a caterpillar leg story to tell us now.

Skills

- ❑ Model multiplication as equal rows (M2).
- ❑ Use language rows *of* and *makes* (M3).
- ❑ Create and solve multiplication story problems (M4).
- ❑ Record multiplication activities using informal word/digit cards (M5).
- ❑ Estimate answers to simple multiplication problems (M8).

How to Make the Activity

- ❑ Copy, laminate, and cut out the four activity cards (pages 70 and 71).
- ❑ Make five copies of the caterpillars (page 72) so you have 10 caterpillars altogether. Color, laminate, and cut out along the dashed lines. These will be ideal for two students to use but can be used by up to five students in a group.
- ❑ Copy the *rows of/makes* cards (page 73). Laminate and cut out along the dotted lines.
- ❑ Place the caterpillars in a storage container and label clearly.
- ❑ Place the informal word cards in a separate container.

Extra Materials Required

- ❑ 60 colorful plastic pegs (in a storage container with lid)
- ❑ two sets of the digits 0–9 from previous activities (page 14) for each student
- ❑ a die for each pair or group
- ❑ 0–9 spinner (page 15) for each pair or group

Variations

- ❑ Demonstrate how to place pegs. Count aloud as you go (e.g., 3, 6, 9, 12). Ask a student to record your story using cards (e.g., four rows of three makes 12).
- ❑ Use more pegs and copies of the caterpillars to explore multiplication facts beyond 60.

Caterpillars   **Exploring Multiplication**

What a Lot of Legs

- ✔ Throw the die.
- ✔ Take that many caterpillars.
- ✔ Spin the spinner.
- ✔ Give each caterpillar that many legs.
- ✔ How many rows of legs?
- ✔ How many legs in each row?
- ✔ How many legs altogether?
- ✔ Find a way to record your actions using the small cards.

How Many Altogether?

- ✔ Make up a caterpillar leg story with a friend.

    e.g., Six caterpillars woke up and discovered they each had four legs.

- ✔ Guess how many legs altogether.
- ✔ Model your story with caterpillars and pegs.
- ✔ Match your number story with cards.

    e.g., | 6 | rows of | 4 | makes | 2 | 4 |

- ✔ Was your guess close?

Caterpillars **Exploring Multiplication**

How Many Rows?

✔ Give some caterpillars an equal number of legs.

✔ Match your actions with cards.

e.g., | 7 | rows of | 3 | makes | 2 | 1 |

✔ Turn over the card that tells you how many rows.

e.g., | | rows of | 3 | makes | 2 | 1 |

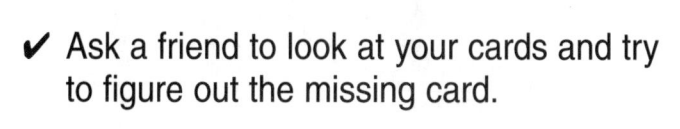

✔ Ask a friend to look at your cards and try to figure out the missing card.

✔ Check then exchange roles.

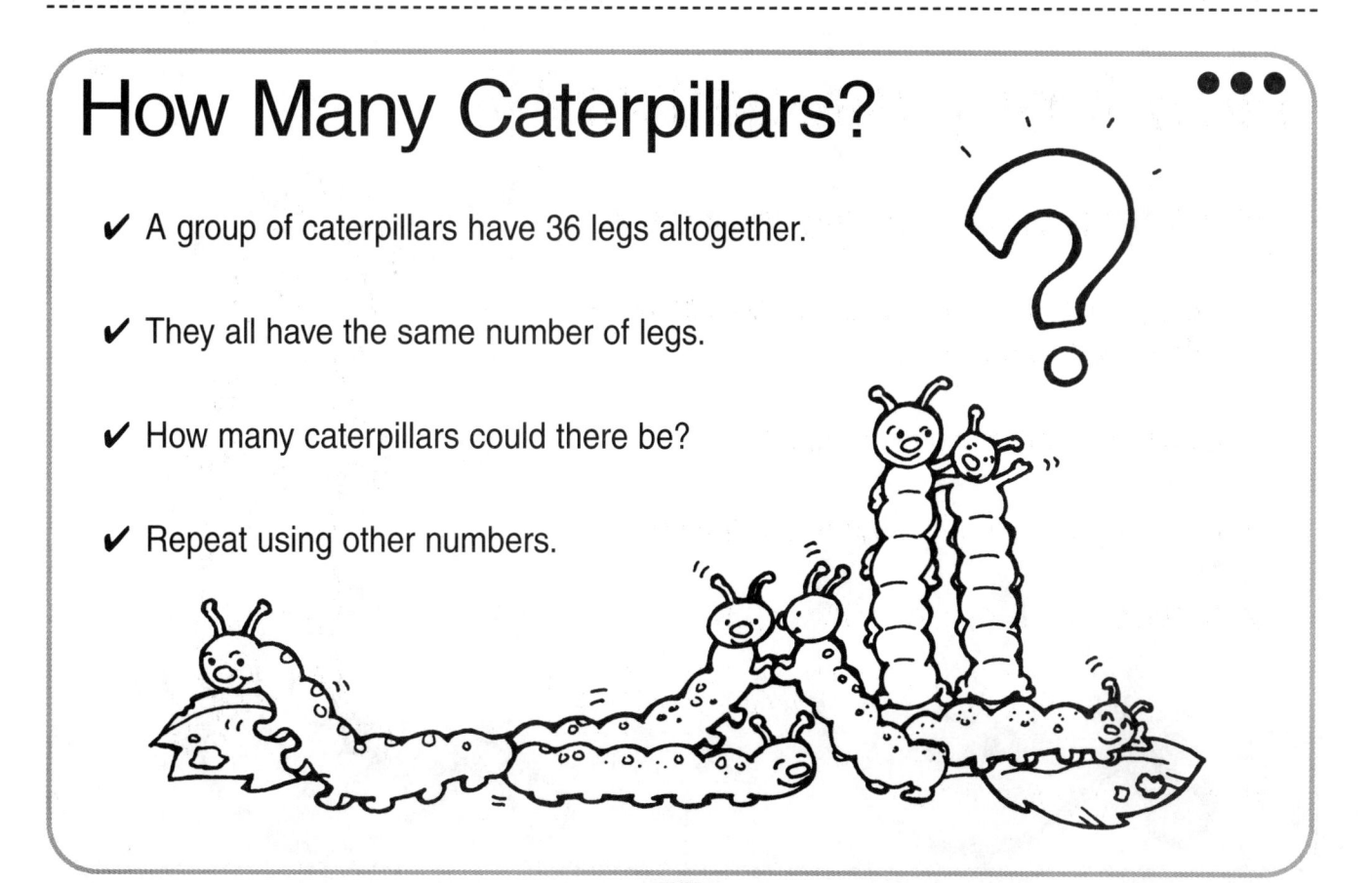

How Many Caterpillars?

✔ A group of caterpillars have 36 legs altogether.

✔ They all have the same number of legs.

✔ How many caterpillars could there be?

✔ Repeat using other numbers.

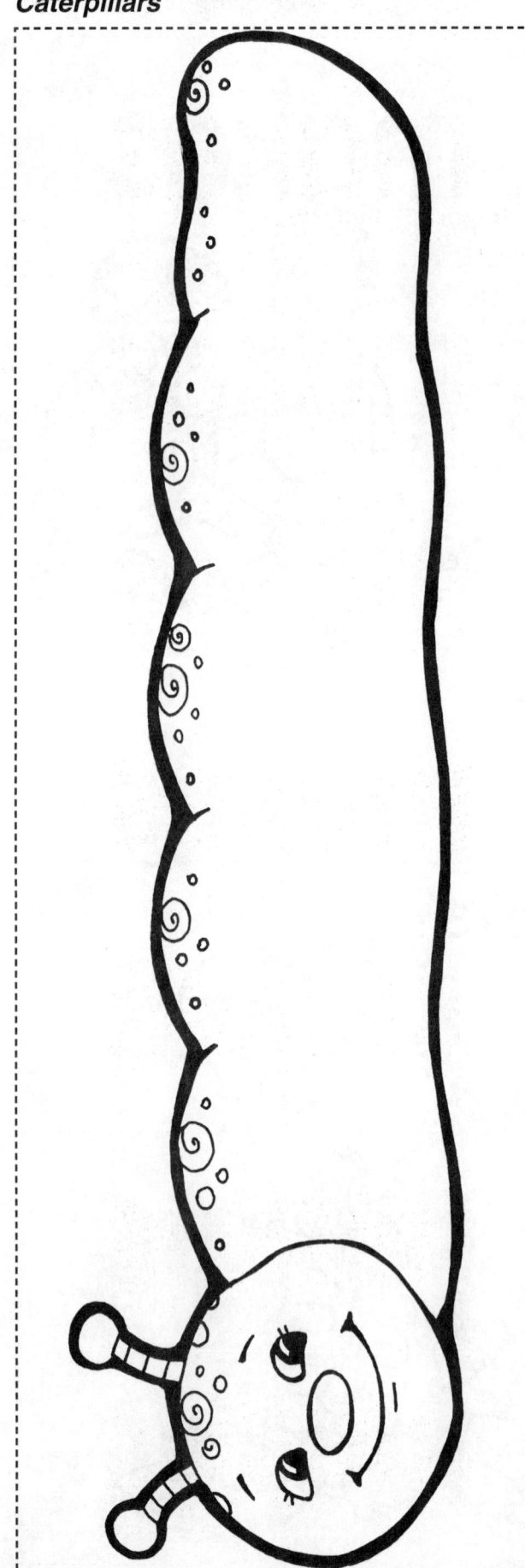

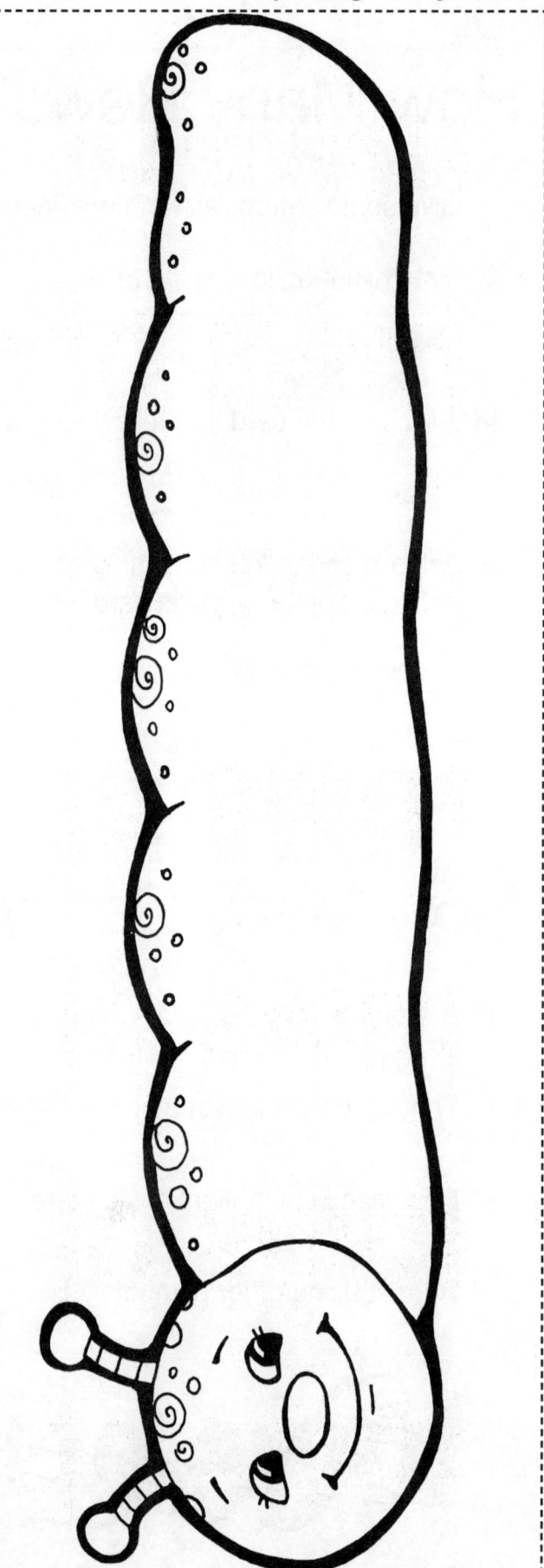

Caterpillars   **Exploring Multiplication**

72

#3526 Math in Action   ©Teacher Created Resources, Inc.

Caterpillars — Exploring Multiplication

rows of	makes
rows of	makes
rows of	makes
rows of	makes
rows of	makes
rows of	makes
rows of	makes

Exploring Multiplication

Smiles

Introducing the Activity

The smiley faces in this activity love to be given away (in paper bags) to people. They are only happy when there is exactly the same number of smiles in each bag. We will discover how many smiles altogether when there are two, three, or more in each bag. We will also practice recording our discoveries using cards for multiplication and equals.

Who can think of a story about smiles to get us started? *(e.g., I would like to give a bag of smiles to three of my friends. If I put four smiles in each bag, how many smiles will there be altogether?)*

How can we model that story using these bags and bottle tops?

How can we use these cards to record our actions? *(Demonstrate.)*

Skills

- Model multiplication as equal groups, rows, or columns (M2).
- Use language—*rows of, groups of, makes,* and *equals* (M3).
- Create and solve multiplication story problems (M4).
- Record multiplication activities using multiplication symbol cards (M6).
- Estimate answers to simple multiplication problems (M8).

How to Make the Activity

- Copy, laminate, and cut out the four activity cards (pages 75 and 76).
- Copy the ⨯ and = cards (page 77). Laminate and cut out along the dashed lines. Store in a separate container.

Extra Materials Required for Each Pair

- 10 paper bags
- 50 smiley face cards (page 78 reproduced on bright-colored heavy stock, if possible, and cut along dashed lines)
- two sets of the digit 0–9 cards from previous activities for each student (page 14)
- You may still want to use the equals word cards (pages 21 and 22) rather than the symbols with some students.

Variations

- Ask some students to stand at the front with a paper bag each. Drop in equal groups of smiley face cards, counting aloud as you go (e.g., 2, 4, 6, 8, 10, 12). Ask a student to write your story using symbols on the chalkboard (e.g., 6 x 2 = 12).
- Use more smiley face cards to explore multiplication facts beyond 50.

#3526 Math in Action ©Teacher Created Resources, Inc.

Smiles — Exploring Multiplication

How Many Smiles?

✔ Work in pairs.

✔ Make up a number story about smiles to tell another pair.

✔ Race to be the first pair to model the story using smiley faces and paper bags.

✔ Use the cards to record your actions.

✔ How many smiley faces are there altogether?

Plenty of Smiles

✔ Put an equal number of smiley faces into some paper bags.

✔ Exchange them with a friend.

✔ How many bags?

✔ How many smiley faces are in each of your bags?

✔ Guess how many smiley faces altogether.

✔ Check and then race your friend to record your discovery using the small cards.

Smiles   Exploring Multiplication

Plenty of Bags

✔ Think of a number between 0 and 10.

✔ Take that many bags.

✔ Think of a number between 0 and 5.

✔ Put that many smiley faces in each bag.

✔ Guess how many smiley faces you have altogether.

✔ Check and then record your actions using the cards.

What a Lot of Bags!

✔ There are 24 smiley faces altogether.

✔ Each bag has the same number of smiley faces.

✔ How many bags could there be?

✔ Repeat using other numbers.

#3526 Math in Action   ©Teacher Created Resources, Inc.

Smiles   **Exploring Multiplication**

×	×	=	=
×	×	=	=
×	×	=	=
×	×	=	=
×	×	=	=
×	×	=	=
×	×	=	=

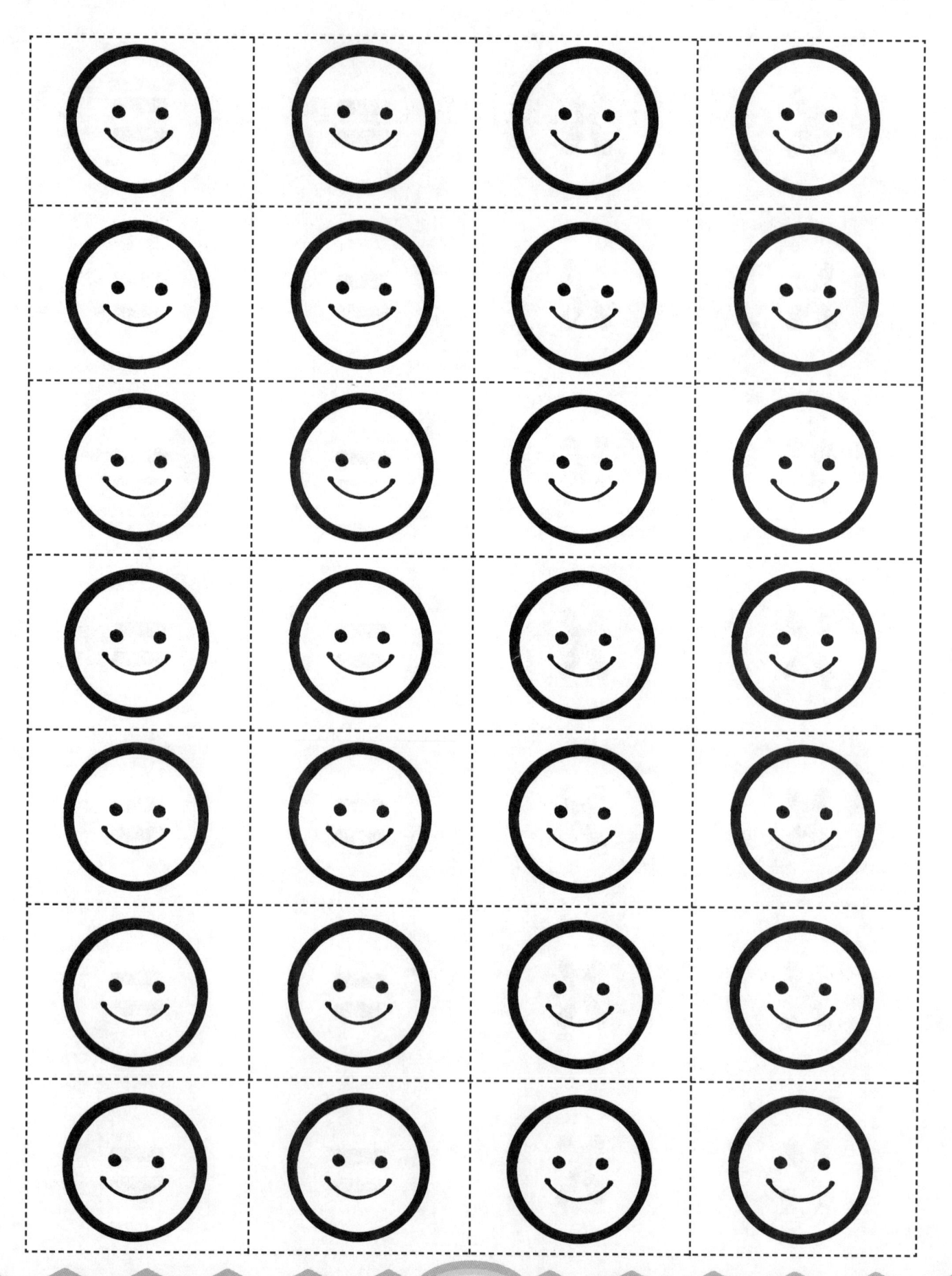

Smiles **Exploring Multiplication**

78

#3526 Math in Action ©Teacher Created Resources, Inc.

Exploring Multiplication

Multiplication Check-Up

Resource
Multiplication Check-Up (page 80)

Activity
The following are suggested instructions and sample comments. Record responses on the Skills Record Sheet (page 95).

"Look at the shapes at the top of your page. What could these be? *(Discuss.)* They are hands but the fingers are missing! Draw five fingers on each hand. How many hands are there? How many fingers altogether? Fill in the missing cards." (M3,4,6)

Sample Profile Comments
Demonstrates and records multiplication actions up to 20, using words and digits.

"Look at the next shapes. What could they be? They are leafs but where are the ladybugs? How many ladybugs should we put on each leaf? How do you know? *(See the number sentence.)* Draw three ladybugs on each leaf. How many leafs altogether? How many ladybugs?"

"Write the rest of the number sentence to match." (M3,4,6)

Sample Profile Comments
Demonstrates and records multiplication actions up to 20 and records using symbols and digits.

"Look at the rows of flying birds. Guess how many altogether. Write your guess on the left. Fill in the missing numbers to make a number story that matches." (M3,4,6,7)

Sample Profile Comments
Recognizes equal rows in multiplication problems and estimates answers before calculating the total number.

"Find the person holding the balloons. Write the number sentence to match. Can you think of any other way to record this? If so, write this underneath (e.g., 7 x 2 = 14)." (M5,6)

Sample Profile Comments
Still unsure of how many groups and the number of objects in each group. We are working on it!

"The spiders at the bottom of your page are missing their legs. How many legs should a spider have? Guess how many legs three spiders will have. Write your guess on the left side of the page by the spiders. Draw the missing legs. Write your own number sentence to match on the line." (M6,7)

Sample Profile Comments
Toula correctly matches her own written number sentence to represent a multiplication action story.

"Turn your page over. In the top half, draw your own picture to show some groups of two. Write your number sentence underneath." (M6,8)

"In the bottom half, make up your own picture story about rows of seven. Record your story with a number sentence." (M6,8)

Sample Profile Comments
Lucy shows confidence when representing multiplication stories in pictures and written number sentences.

Multiplication Check-Up

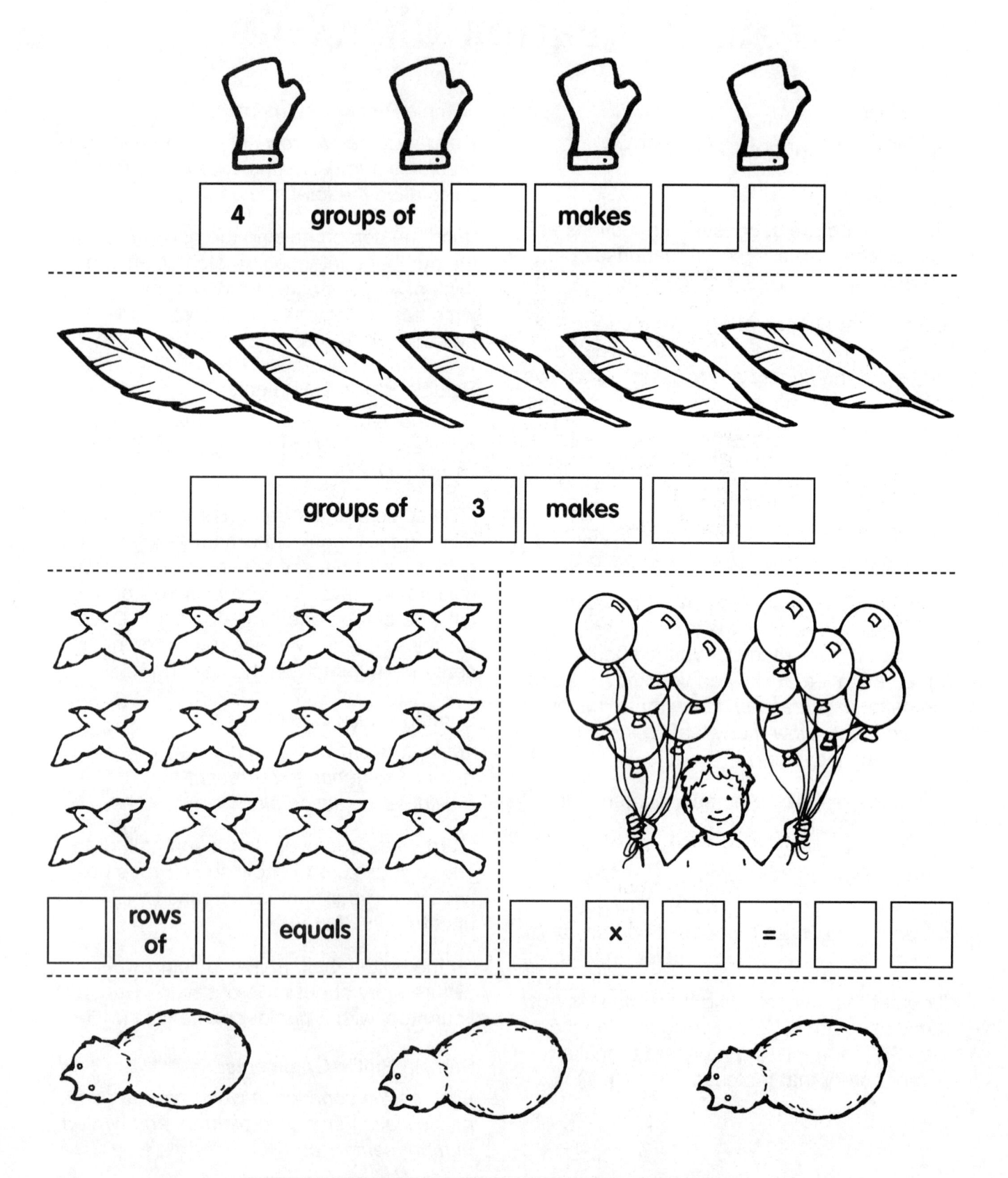

4 | groups of | | makes | |

| | groups of | 3 | makes | |

| | rows of | | equals | |

| | × | | = | |

Exploring Division

In this unit, your students will do the following:

- ❑ Recognize shares as "fair" or "unfair."
- ❑ Divide a set of objects by sharing or grouping.
- ❑ Understand that sometimes there are objects left over.
- ❑ Model division story problems using concrete materials.
- ❑ Use informal language to describe division actions.
- ❑ Understand that division is just repeated subtraction.

Division "Active-ities"

Handfuls PAIR D1,2,5

Resources
up to 20 plastic counters (e.g., dinosaurs, teddy bears, fish, or frogs available from educational suppliers) for each pair

Activity
Work with a partner. Grab a handful of counters. Guess whether there will be a fair share. Check by sharing the counters between the two of you. What will you do with any leftovers? Repeat.

Variations
Share more than 20 counters.

Work with 3, 4, or 5 people in each team.

Smarties PAIR D1,2

Resources
20 plastic counters, 9 empty matchboxes, and a set of 0–9 digit cards (page 14) for each pair

Activity
Place the digit cards face down in the center. Take a handful of counters. Count them. Turn over a digit card and take that many matchboxes. Share the counters between those matchboxes. Was there a fair share?

Pasta GROUP D1,2,5

Resources
a margarine container filled with 20 pasta pieces for each group

Activity
In turns, take a handful of pasta and guess whether you will all get a fair share.

Check by sharing the pasta with your friends. Does everyone get a fair share?

How many pieces does each person get? Were there any pieces extra?

Return the pasta to the container and then repeat by taking another handful.

Variations
Share more than 20 pieces of pasta.

Teams CLASS D1,2

Activity
Work outside. Select 2–10 students to be group leaders. Ask the other students to share themselves equally among the teams quickly. Can they form equal teams in less than two minutes? Do all teams have the same number of students? Are any students left over? Decide on an activity for the teams to do together (e.g., race to the fence and back).

Exploring Division

Pirates

Introducing the Activity

These pirates love to hunt for hidden treasure, but they really squabble if any pirate ship has more pirates on it than another. The pirate captain is always careful to make sure that there is a fair share of pirates on each ship. Sometimes the captain has to rescue pirates stranded on an island.

Who can tell us a story about these pirates?

    (e.g., The pirate captain had two ships and seven pirates.

    How many pirates can sail on each ship?)

Explain how to model this story using the pirates and the pirate ships.
(There will be one extra pirate who stays behind with the captain.)

Skills

- ❑ Recognize equal and unequal shares (D1).
- ❑ Divide a set of objects by sharing (D2).
- ❑ Create division story problems (D4).
- ❑ Estimate answers to simple division problems (D5).

How to Make the Activity

- ❑ Copy, laminate, and cut out the four activity cards (pages 84 and 85).
- ❑ Make ten copies of the pirate ship (page 86). Color and laminate.
- ❑ Make up to six copies of the pirates (page 87). Color, laminate, and cut out. Store in a separate container with a lid.

Extra Materials Required

- ❑ 0–9 spinner (page 15)

Variation

- ❑ Make up your own pirate stories about equal shares.

Pirates   **Exploring Division**

Ship Ahoy

- ✔ Take a handful of pirates and a few ships.

- ✔ Guess how many pirates and then count.

- ✔ Share the pirates between the ships.

- ✔ Is there a fair share?

- ✔ What will you do with any extra pirates?

Rescue Me

- ✔ There are 30 pirates stranded on an island.

- ✔ Spin the spinner to see how many ships the captain has available to rescue them.

- ✔ Guess if there will be a fair share and then check.

- ✔ How many pirates will go in each ship?

- ✔ Are any pirates left behind?

#3526 Math in Action   ©Teacher Created Resources, Inc.

Pirates   **Exploring Division**

How Many to the Rescue?

✔ The pirate captain sent some ships to rescue 15 pirates.

✔ There was more than one ship and there were no pirates left behind.

✔ How many ships could have been sent?

Stranded

✔ Today the pirate captain has 20 pirates who want to find treasure.

✔ There are up to five ships available.

✔ How many different ways can the captain send the pirates off to discover treasure?

Pirates **Exploring Division**

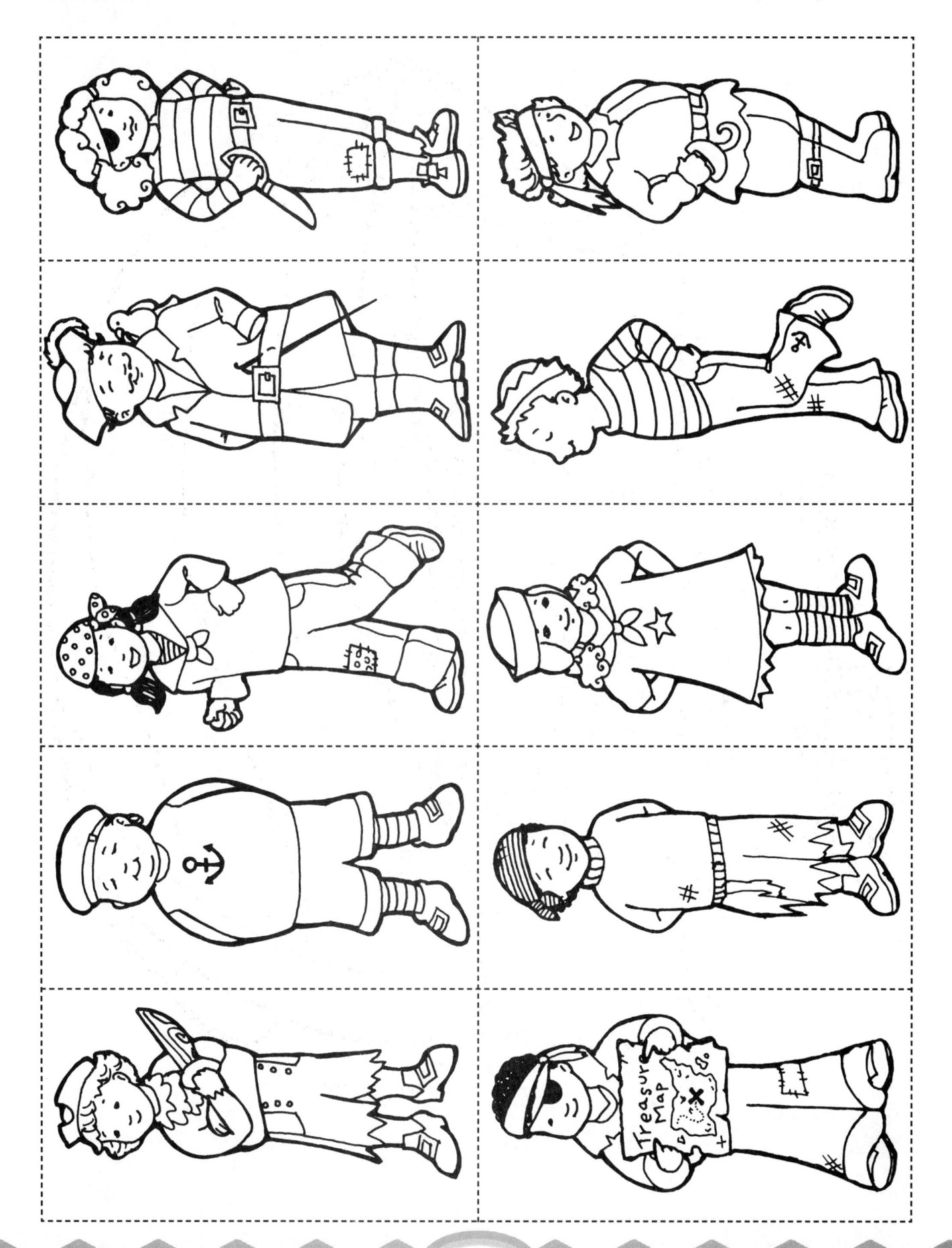

Pirates

Exploring Division

87

©Teacher Created Resources, Inc.

#3526 Math in Action

Division Review

Hoops `CLASS` `D1,3`

Resources

up to 30 hoops in a pile to one side

Activity

Work with the whole class, preferably outdoors. Call out a number 0–5. Students race to form groups to match that number, take a hoop, and then stand together inside it. Is there now an equal number of children inside each hoop? Are any students left over?

Variations

Call out numbers 0–10 and omit the hoops. Form huddles by placing your arms around each other to signify each group.

Fish `GROUP` `D1,3`

Resources

20 plastic fish counters (or other counters), 10 small yogurt containers with lids (these are the aquariums), and a 0–9 spinner (page 15) per group

Activity

In turn, take a handful of fish and ask your friends to guess how many fish altogether. Check by counting. Take turns to spin the spinner. This number tells you how many fish to place in each aquarium. How many aquariums did you need? Were any fish left over? What will you do with the extra fish? *(e.g., put more in one container, buy more fish, etc.)*

Can you make up a story to match these actions?

Variations

Use more than 20 fish counters.

Spaceship `PAIR` `D1,3,5`

Resources

interlocking building blocks, up to 20 plastic/wooden people figurines

Activity

Build a spaceship to hold three people. If there are 15 astronauts who want to go to the moon, how many different trips will you need to make? Guess first and then model using your spaceship and the tiny people. Were any astronauts left over?

Variations

Vary the total number of astronauts wanting to use your spaceship each time (e.g., build a spaceship to hold 2, 4, 5, or more astronauts).

Division Review

Funny Faces GROUP D1,3,4

Resources

plain oval cookies, icing mixture in a bowl, suitable decorations (carrot slivers, raisins, and jellybeans), plastic knife for spreading the icing

Activity

Ice at least one cookie per student in the group. (*Note:* Check for food allergies and sugar sensitivities.) Take a handful of raisins.

Place two of these as eyes onto each face. Can every face get two eyes?

Take 30 carrot slivers. Is that enough for each face to get five strands of hair?

Take seven jellybeans. Is that enough for each face to get a big smile?

Invent your own sharing questions related to the decorations available.

Farm Animals GROUP D1,3,4

Resources

50 or more plastic farm animals, fence materials

Activity

Make up to 10 enclosed areas from the fencing material. If each enclosed area gets three animals, how many enclosed areas will you need for 13 animals? 18 animals? 25 animals? What if each paddock holds five animals? Make up your own grouping problems, too.

Scarecrows GROUP D1,3

Resources

10 copies of the scarecrow (page 90) colored and then laminated; three copies of the birds (page 91) colored, laminated, and cut out as individual birds; fair share cards (page 92) laminated and then cut out; 0–9 spinner (page 15)

Activity A

Work in a group of up to six students with a scarecrow each and the pile of birds in the center of the group. In turn, take a handful of birds and ask everyone to guess how many there are altogether. Spin the spinner. The number tells you how many birds to put around each scarecrow. How many scarecrows did the birds fly around? Was there an equal or an unequal share? Are there any birds left over? What will you do with these?

Activity B

Place the fair share cards face down in a pile in the center. Turn over the top card. Use the scarecrows and the birds to make up a scarecrow story to match this card. Use the ? card to make up your own question (e.g., an unequal share: There were six birds that wanted to annoy two scarecrows. Four birds flew around one scarecrow and two birds flew around the other one.). Once you have told your story, discuss with your friends how to model it using the birds and scarecrows.

Scarecrows **Exploring Division**

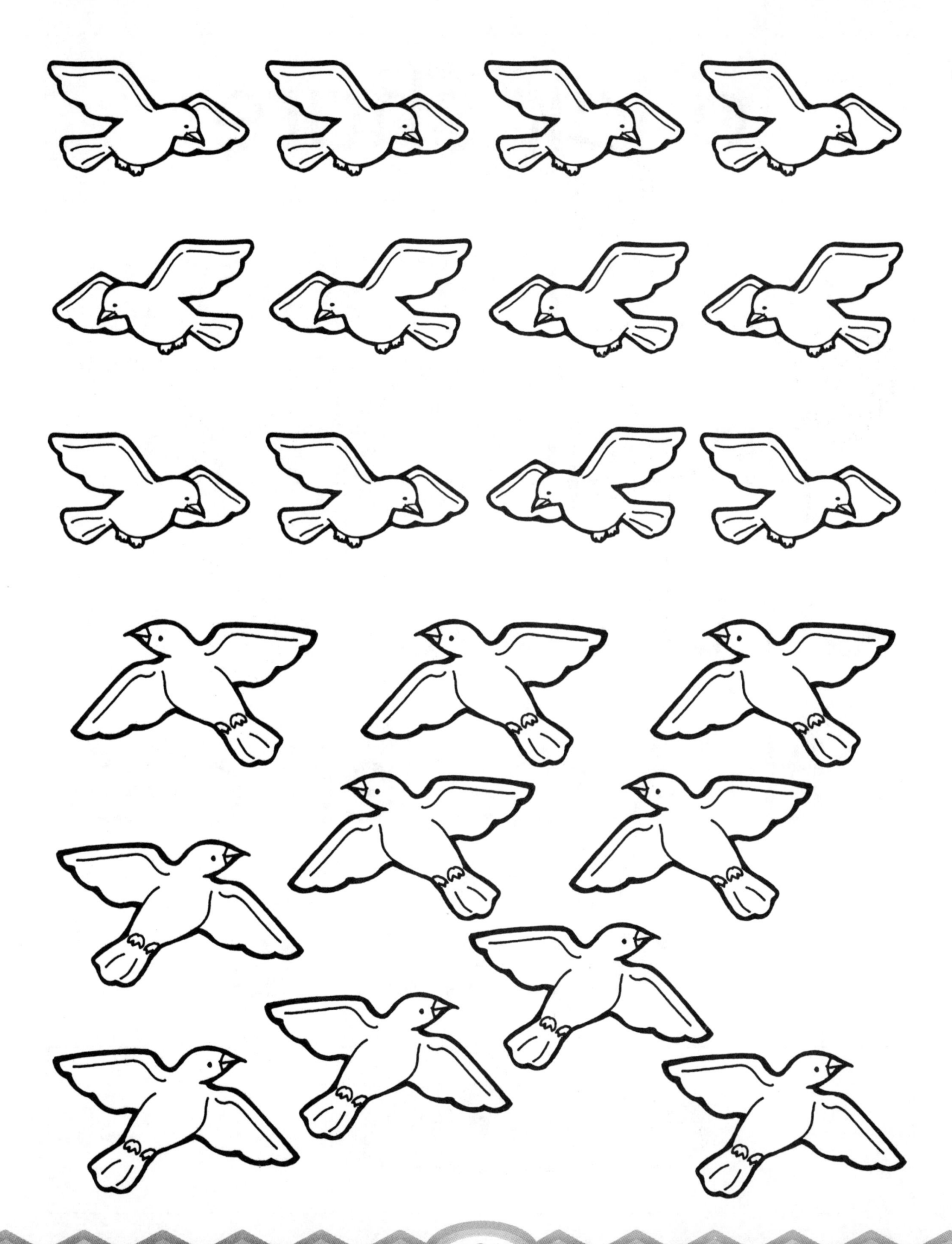

Scarecrows **Exploring Division**

a fair share

not a fair share

an equal share

an unequal share

with some left over

with extras

?

Exploring Division

Division Check-Up

Resource

Division Check-Up (page 94)

Activity

The following are suggested instructions and sample comments. Record responses on the Skills Record Sheet (page 95).

"Look at the bowls of fruit. How many pieces are in each bowl? Are there equal numbers in each bowl? If not, draw in the extra fruit so that each bowl has an equal amount." (D1)

Sample Profile Comments

Recognizes when groups are unequal and combines objects to form equal groups.

"Find the three duck ponds under the bowls of fruit. How can you share seven ducks equally among the ponds? Draw your solution. If there are any left over, draw them on the right." (D2)

Sample Profile Comments

Understands that sharing means giving out one item at a time to each group.

"Look at all the animals that live on Mrs. Brown's farm. She wants them put into groups of six so she can check how healthy they are. How many groups of six will there be? How can you show this? *(e.g., Draw a line around each group of six faces.)* Put a cross on any animals left over." (D3)

Sample Profile Comments

Divides a large set of objects by placing into smaller groups.

"At the bottom of your page, there are five boxes. What could we put into these? *(e.g., balls)* Each box holds exactly four balls. If there are 16 balls, how many boxes will we need? Show me by drawing balls in the boxes. Are there any balls left over?" (D3)

Sample Profile Comments

Understands how to solve simple division problems by drawing a given number of groups.

"Turn your page over. In the top half, draw your own picture to show how to share some objects among three people." (D2,4)

"In the bottom half, draw your own picture story about groups of ten." (D3,4)

Sample Profile Comments

Distinguishes sharing and grouping actions in simple division problems.

©Teacher Created Resources, Inc. #3526 Math in Action

Division Check-Up

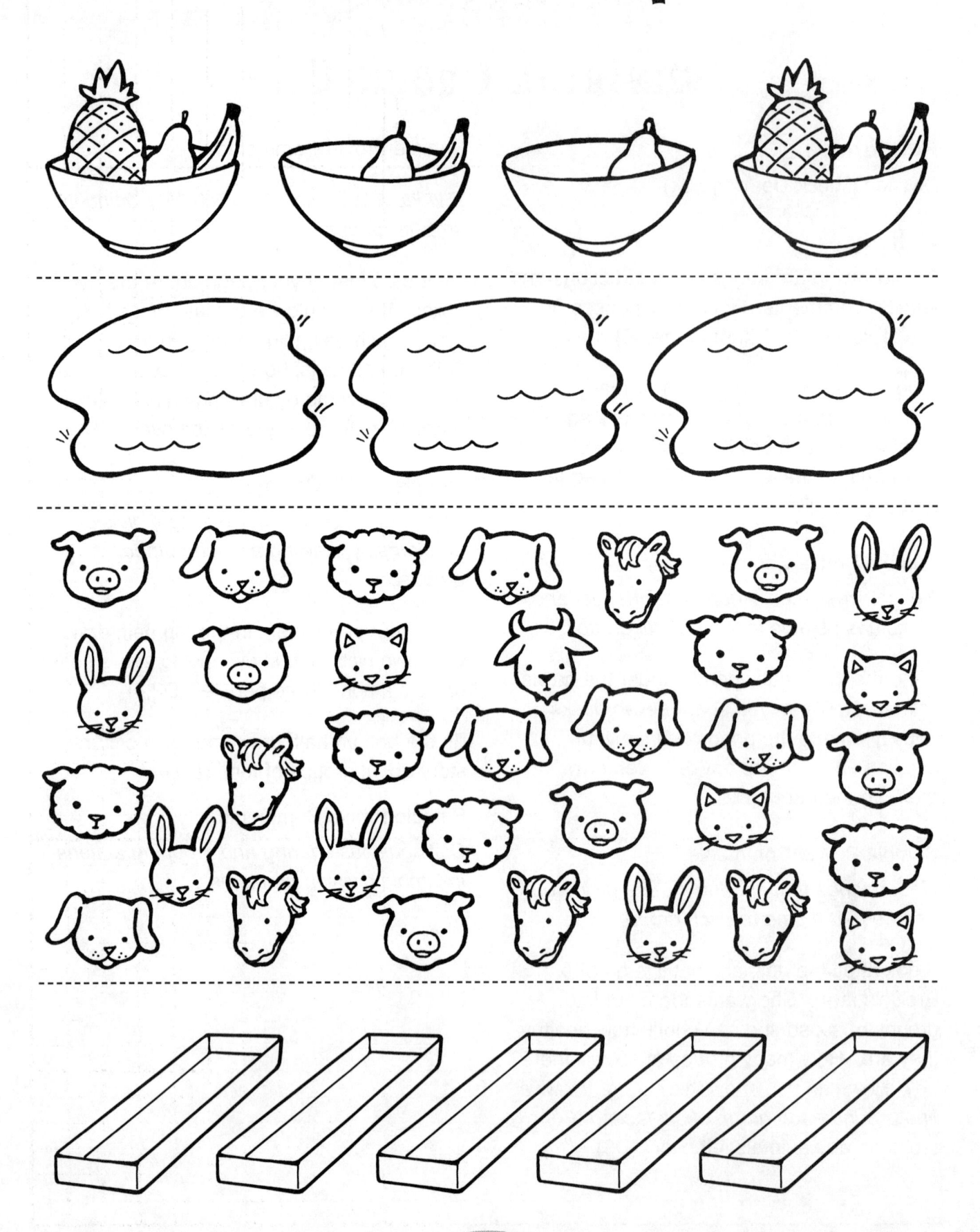

Skills Record Sheet

MATH IN ACTION: Operations Activities 0–50

Reproduce copies as needed to accommodate class size.

NAME

Category	Code	Skill
Adding	A1	Demonstrates addition facts to 10 with objects
	A2	Demonstrates addition facts to 20 with objects
	A3	Creates, solves addition story problems
	A4	Records addition activities using word/digit cards
	A5	Records addition activities using symbol/digit cards
	A6	Records addition activities using written number sentences
	A7	Estimates answers to addition problems
	A8	Uses a number line to solve addition problems
	A9	Recalls addition facts to 10 using a range of strategies
	A10	Recalls addition facts to 20 using a range of strategies
	A11	Adds three or more digits up to 20
	A12	Adds to 50 using objects
Subtracting	S1	Demonstrates subtraction facts to 10 with objects
	S2	Demonstrates subtraction facts to 20 with objects
	S3	Creates and solves subtraction story problems
	S4	Records subtraction activities using word/digit cards
	S5	Records subtraction activities using symbol/digit cards
	S6	Records subtraction activities using written number sentences
	S7	Estimates answers to subtraction problems
	S8	Uses a number line to solve subtraction problems
	S9	Recalls subtraction facts to 10 using a range of strategies
	S10	Recalls subtraction facts to 20 using a range of strategies
	S11	Subtracts 3 or more digits from numbers up to 20
	S12	Subtracts from 50 using objects
Multiplying	M1	Counts by 2s, 5s, and 10s to 50 or more
	M2	Models multiplication as equal groups, rows, or columns
	M3	Uses language "groups of," "rows of," or "makes"
	M4	Creates and solves multiplication story problems
	M5	Records multiplication activities using word/digit cards
	M6	Records multiplication activities using symbol/digit cards
	M7	Records multiplication activities using number sentences
	M8	Estimates answers to simple multiplication problems
Dividing	D1	Recognizes equal, fair/unequal, or not fair shares
	D2	Divides a set of objects by sharing
	D3	Divides a set of objects by grouping
	D4	Creates and solves his or her own division story problems
	D5	Estimates answers to simple division story problems

©Teacher Created Resources, Inc.

Sample Weekly Lesson Plan

STRAND Number
GRADE 1
SUBSTRAND Exploring Multiplication Facts to 20
TERM 2 **WEEK** 7

LANGUAGE
- "equal groups of" and "equal rows of"
- "There are ___ groups altogether."
- "There are ___ objects altogether."
- "That makes ___ altogether."
- "That equals ___ altogether."

SKILLS
- model and describe equal groups or rows of objects
- label the number of groups or rows
- label the number of objects in a row or group
- find and label the total number of objects in the groups
- create and solve multiplication story problems

RESOURCES
- Number Roll-Ups (page 61)
- counters (e.g., frogs, bears, shells dinosaurs, or fish)
- Digit Cards 0–9 (page 14)
- label cards "groups of" and "makes" (page 68)
- label cards "rows of" and "makes" (page 73)
- calculators
- dog and dog bones (pages 66–67)
- Caterpillars (page 72)
- plastic pegs, scrap paper, pencils

MONDAY	TUESDAY	WEDNESDAY	THURSDAY	FRIDAY
• Review counting by 2s, 5s, and 10s. Demonstrate with children (counting eyes, fingers, or toes). • Model groups of 2s, 5s, and 10s with counters. How many groups? How many in each group? How many altogether? • Make Number Roll-Ups for 2s, 5s, and 10s. Count forwards/backwards. • Whole class problem—"I have 4 dogs. Each dog has 2 bones. How many bones?" Invent your own problems for homework.	• Check homework. • Practice counting with Number Roll-Ups. • Review "groups of." Invent stories to model. Demonstrate how to use label cards. • Group tasks—counters, label cards (Rotate desks after two oral instructions e.g., "Show me three groups of two bears.") • Outdoor games: Bunch Up and Exercises	• Introduce game "Rows." • Demonstrate making rows with counters to match stories. • Demonstrate how to check answers with a calculator. • Group tasks—counters, label cards (Rotate desks after two oral instructions e.g., "Show me five rows of two frogs.") • Finish with class challenges (e.g., "I had five cats with two kittens each. How many kittens altogether?").	• Demonstrate "Dog Bones" and "Caterpillars." • Free exploration and discussion in groups • Observe and record individual responses (See Skills Record Sheet page 95.). • Finish with class challenges (e.g., "I saw three rows of cars. There were four cars in each row. How many cars altogether?" "How can we record this on the board? Who has another way?"	• Review multiplication activities. Discuss what they liked best/least. • Review using words and digits to record multiplication stories. • Whole class: Draw what you hear (e.g., three groups of five balls). Write the number sentence underneath. • Outdoor games: Bunch Up, Rows, and Exercises.

#3526 Math in Action 96 ©Teacher Created Resources, Inc.